FOOD:
FEAR, FAITH & FELLOWSHIP

An Interdisciplinary Examination of
Food Systems in the United States

VANESSA KOVAROVIC

Food: Fear, Faith & Fellowship

© 2022 Vanessa Kovarovic

ISBN 978-1-66785-405-2
eBook ISBN 978-1-66785-406-9

For Lisa.

A long overdue thank you for that made from scratch cherry pie that I didn't have the capacity to appreciate in the moment.

Thank you

There are so many people who provided nudges, encouragement, feedback, perspective, accountability, and patience while I was working on this project. The journey began years ago even though the book didn't become a reality until much later. Barbara Chatzkel, thank you for providing the initial wind in my sails. The educators at Empire State College including but not limited to Linda Jones, Lorraine Lander, Anastasia Pratt, Himanee Gupta-Carlson, Karyn Pilgrim, Kim Stote, Anna Bates, Cathy Davison, Menoukha Case, and Duncan RyanMann whose professional expertise combined with a desire to increase knowledge by promoting critical thinking helped me to stay focused when the going got tough. To my closest friends, my brief acquaintances, and everyone in between who indulged my questions and discussed new perspectives over countless meals, your contributions keep this book real and accessible to any reader. Most importantly, my endless gratitude to Mike Hands whose proofreading is only a small part of his overall support on this journey. I could not have written this book without him.

TABLE OF CONTENTS

INTRODUCTION

This book began with a list of questions I had regarding our food systems. The more I read, pondered, and discussed my questions with friends, co-workers, and professors, the more questions I had. As my list grew, I read and researched further, and the questions kept coming. The complexity of food systems in the United States is both a marvel and a mystery. I discovered a plethora of really good books about food written from a wide variety of perspectives, many that you will find cited throughout this text. I have chosen all the sources used here carefully, looking for evidence that supports opinions and calling out those that don't back up their claims.

Many of the sources used challenged my beliefs and caused me to think deeply about my mindset with regards to food. In order to comprehend the larger conglomeration of related points, I did what most scientists do—I broke it down into bite-sized pieces. Considering food from a specific viewpoint helped to identify critical points that impact our food security with regards to that perspective. I would then begin to think about practices that might address the issue I had identified, only to realize that the solution posed might negatively impact our food systems when considered from a different viewpoint, or simply wouldn't be feasible in reality even though it made sense on paper.

A common example of this type of conflict would be looking at food from the health perspective and acknowledging that our diets simply must change to improve the health of the individual and, by extension, the health of the workforce and reducing the portion of the economy spent to care for those who are sick. This is a simple and basic idea. One might then consider many options for improving health through education and access to fresh whole foods, and how those efforts can trigger a shift in the way the population eats. Yet evidence shows that the overall improvement in diet because of education and access impacts only a limited segment of the population and tends not to be long lasting, despite the benefits.

To understand why, one must consider from the political perspective how subsidies from the government are targeted at supporting mass production of grains that become processed foods, and how limited regulation of the advertising industry meant to protect a free market allows for people to be bombarded with "eat more" messages as biologist and nutritionist Marion Nestle calls them, result in an overall food culture that does not support health as a primary goal. The influences of marketing and plentiful low-cost imitation food are overwhelmingly more pervasive than eat healthy messages and distract consumers from what is best to eat. Additionally, from the economic perspective, the ongoing health crisis feeds a large sector that offers products claiming to improve health and well-being. Promoting businesses and encouraging better health simply don't have a shared end goal; if half of these products worked as promised, they would have a shrinking customer base. In the end, it is not a surprise that consumers are getting sicker instead of healthier, and the industry offering solutions grows rather than dwindles. There is no incentive for businesses to reduce the number of customers who might buy their products by promoting a whole foods diet as the best solution for health issues.

This example demonstrates how important it is to have a sound understanding of the various perspectives or "disciplines" that have a stake in our food systems, and to consider how they influence one another. Stating the importance of an interdisciplinary approach as an idea is easy to type on this page, but biting into the meat of it (food pun intended) feels a lot like trying to run through a shallow pond lined with thick mud that pulls your shoes off and submerged tree trunks that are just waiting to trip you up and send you face first into the murky water. There were many times in the past few years when trying to comprehend our food systems felt incredibly frustrating, and several of my friends questioned "why don't you just ride a floaty across the pond instead of dealing with all the hidden obstacles under the surface?" That sentiment helped me understand something very important. I realized how much cultural influences have degraded our perception of food security and encouraged us to coast rather than be engaged. It is easy for people to find something to float on and ignore that our food systems are becoming less and less secure each season as long as the grocery store is available twenty-four hours a day seven days a week in the present moment. Only when there is a localized crisis, and the food system fails "without warning" does the public wake up and begin looking for who is responsible without considering how each consumer's actions set the stage for exactly this kind of situation. Wendell Berry has written extensively about this modern cultural attitude.

From my experience, no one wants to be told to add one more item to their three most important things to do today, and what should be bumped so food can take a top spot? There are so many other things that demand our attention day to day: working to make a living or trying to find a job or living on a fixed income, taking care of children and other family members, dealing with political issues, trying to navigate through the global COVID-19 pandemic, and trying to find a few minutes to relax and enjoy being with friends and family. Still, these daily concerns are all related to our food systems in overt ways, and our urge to ignore connections doesn't change the fact that we each have to eat to live. It also leaves no room to visualize food as a solution to daily stresses, and that is part of what has driven me to push through the frustration. Our food systems are fragile and that is terrifying, but each person has the ability to improve the food systems they are fed by in real time and, as a result, improve their entire life experience. Focusing attention on our food choices doesn't have to be a chore; for me, it has been a way to put the rest of my life in perspective rather than allowing myself to become overwhelmed by my day to day.

I understand there are dramatic differences between the resources each individual has access to in order to make changes in what food they eat. I have experienced periods in my life where lack of money influenced every choice I made, including sometimes buying food instead of paying a bill and eating the cheapest food I could buy. When I think back to those moments, I can easily imagine how angry I would feel if someone told me to spend more on food. From my research, I can now see that our society's fixation on the cost of food is part of what drives the increase in food insecurity, and that eating more whole fresh foods does not necessarily mean spending more. It means changing what we choose to eat, for example learning what is "in season," buying, and preserving whole foods at their freshest and least expensive. It also means recognizing that changing food habits have a direct influence on what we spend for health costs; shifting where we spend the money we have is different than simply spending "more." Nearly every person has some ability to make small changes in their eating habits, and if enough people strive to eat differently, it will trigger a change in our food systems as a whole and open up opportunities to our least fortunate neighbors. This idea is not about charity, it's about realizing that we are more secure when our neighbor is also secure. Making adjustments to how we view food and how what we eat impacts our bodies and minds, our relationships, our community, and our economy. It is the base for everything else in our lives; no other action has such far reaching influence because eating is the one action we must all do regularly to survive. As you continue to read, I hope you will find I have made every attempt to suggest changes that can be applied by any reader regardless of their financial status. Ultimately, the most important step we can all take, thinking about food differently, costs nothing.

Returning to my recent journey to better understand food systems from various perspectives, I searched for resources that used an interdisciplinary approach to examine the country's food systems. While there are many brilliant food experts who have published resources, including some who occasionally point out interdisciplinary connections, I could not find any texts that purposely explored how various disciplines form conflicting or supportive relationships with each other. In the scientific research community, the idea of an interdisciplinary approach exists but isn't prevalent, despite how it has been shown to help shed light on and even test new hypotheses more thoroughly. With regards to food systems, I found that despite how frustrating it can be, dealing with the complexity is necessary, even crucial to reducing insecurity. Like every other situation we face in our daily lives, grabbing that floatie and ignoring warning signs about turbulence up ahead only leads to a bigger problem in the future. Psychologically, it also increases worry and fear, and those emotions tend to cause us to delay addressing the problem even longer, compounding the negative.

During my research and struggle, the idea for this book was born. I couldn't find the interdisciplinary resource I needed to guide me, which meant if someone else started down the same path I was on, we would each be working from scratch and independently. Dialog and debate are important tools that bring value to a topic as complex as this; someone else always has a new observation or way to look at a situation. That idea is why I began creating an interdisciplinary resource that may shine a light that invites others to join me in looking for solutions and encourages everyone to take part in the discussion. Trying to cross that pond with the mud and the tree trunks is a lot easier and even fun with companions.

The value of working together to consider food insecurity issues in depth and striving to understand the variety of system connections as they relate to each other goes beyond tapping into the power of diverse viewpoints. It helps us realize that this problem is universal. All humans are impacted. All humans must eat to survive. If our neighbor feels the strain of food insecurity before we do, it is a warning signal of trouble for all of us, not simply a result of one person's situation versus another. Financial means may insulate some from being as aware of food system frailty and may provide others the ability to react to some unexpected situations, but in the end, a severe weather event or other unforeseen catastrophe will be felt by everyone regardless of wealth. Money can buy access to many things, but it can't improve soil health. Money can pay for the medical treatments needed when our foods, contaminated with chemicals, cause our bodies to fail. But wouldn't it be better to not get sick in the first place? Paying more to compensate for food insecurity shortfalls is possible in the short term, but like the resources used to grow and distribute our food, money is a resource that can disappear or lose value with little warning.

I figured out early on in this project that I really needed to see food through other people's eyes, because my knowledgebase was growly so rapidly that it was hard to consider alternative perspectives. In addition to the resources I was reading, I realized I needed feedback from real people to form a comprehensive picture about food. The experts have spent a great deal of time doing their own research, and food is a huge part of their lives, not just at mealtime. Food was on my mind constantly, and I could see a gap developing between my thought processes and those of the people closest to me. I needed to understand a baseline of a wide range of habits and traditions from a diverse group of people in order to keep this book relevant. Beginning in the spring of 2017, I created a collection of "edible landscape gardens" all around my property.

From the summer of 2017 to until the COVID-19 pandemic began to impact my state in March of 2020, I committed to hosting at least two to three casual small dinner parties a month, inviting a wide range of people from different parts of my social network. During these parties, we physically spent time immersed in these gardens. I asked my guests to help harvest ingredients for our meals, when available, and I asked many questions about each individual's food culture. I shared knowledge when asked, but primarily, these dinners were meant for me to better understand my guests' current attitudes toward food rather than change their perceptions. The atmosphere was kept light and welcoming to ensure guests felt at ease with honestly sharing their food cultures.

These dinners helped keep me grounded. My goal for writing this book has never been to become a food expert; there are plenty of very intelligent people who deserve that title already. My goal was to draw connections between all the information out there so that everyone who eats in order to maintain their physical life might pause to consider, and then take steps to improve their relationship with food and the systems that provide that food to them. I strongly believe everyone has a stake in the frailty of our food systems whether they realize it or not, and everyone can take some action to improve their food practices regardless of whether their motives are focused on their own benefit or if they are community minded. Recognizing the importance of connecting with food in a different way and learning how to visualize connections across disciplines shouldn't just be left to the experts. We need them to provide us detailed insight because we ourselves are experts in other areas and may not have the time to dedicate to research, but we all need to be able to look at food as something beyond just a "thing" like all the other things in our life. Food is different. My life would be really hard if I didn't have a car so I am likely to say I "need" one, but I could adapt and use another form of transportation if it came down to it. Alternatively, I need food to survive. If I don't eat, my body will begin to fail within days, and I will die. My need for food is different than my need for a car. It is that nuance that we all must relearn.

Ultimately, this book is about recognizing connections and their impacts on our individual lives. The under-acknowledgment and disregard of this point is dangerous: ignoring or being blind to connections erodes the potential each person has for enjoyment, satisfaction, and creativity in their life experience. I have friends who would question if life should be a pleasurable journey, suggesting that this moment we are in is not as important as whatever comes next. That mentality is a sign of sickness in my opinion. Regardless of what we believe happens to us when our human experience ends, whether it's heaven, reincarnation, or nothingness, dismissing the present moment demonstrates a lack of perceived value in our existence. Believing that there is a purpose, reason, or value to our place in the global ecosystem (Gaia) is fundamental to all other belief systems. The alternative viewpoint, that humans have no value and exist separately from the ecosystem surrounding us, hints at a crisis or illness in our identity or soul. While historically, humans have been held up as more advanced than other living beings, I wonder if our ability to emotionally dismiss our value and by extension our power and responsibility is a quality that makes us less advanced than the living beings that embrace their roles within the greater ecosystem with consistency and dedication. Even worse is the trend of using domination to fill the void left by a lack in self-valuing, which usually results in an abuse of power.

Culture plays a huge role in guiding our self-defined image, but the fact that we can be manipulated to forget that as individuals we have worth and are alive in this moment with the ability to experience joy, gratitude, and empathy is a detriment to our species. Unbalanced emotions tilted toward the negative lead to despair, torment, and depression and make it difficult to move forward in a productive way. Without a supportive community that re-enforces the value of the individual and the potential for creative expression that highlights that value, our species become self-destructive. Our destructive actions are not just limited to our species though, and that is a prime clue that we aren't separate from the ecosystem even if we want to believe we are. Our lack of care for our own life experience and dismissal of our impact on the world around us dooms our entire living planet. Conversely, finding a way to renew our mindset and value the life energy given to us through food can inspire us to greater heights than we may be able to conceive of at this time.

My hope is that in reading this book, you will discover the following: a new respect and appreciation for food, yourself, and your community. At the very least, I wish that each reader learns something about food system connections that helps enlighten their perception and influences their choices going forward. The point of this book isn't to tell the reader what to eat, rather to challenge them to think about why they make the choices they do. I have been mindful of the potential for this text to trigger emotional reactions when the reader is confronted with an idea different than what they believe. The possibility has been a concern because once someone has been emotionally triggered, they often stop interacting or absorbing new information. For this reason, I have cited this book extensively so that information that conflicts with an individual's world view can be further explored and the dialog can continue rather than stall. I ask that if something triggers you, please take a moment to consider what you are feeling or reacting to. Be aware that feeling defensive may be natural but can limit engagement. The topic of food and how it impacts our survival demands that we all participate in protecting our food systems. Every perspective has value if it is presented calmly with respect. Thank you in advance for spending your valuable time reading this book.

CHAPTER 1

DEFINING FOOD, FOOD SECURITY, AND FOOD SUSTAINABILITY

How to start this first chapter has been bouncing around in my head for months. It is clear to me that I must explicitly define, for you the reader, up front what I understand certain words and concepts to mean for several reasons. Terms related to food are overused, still developing, or are not always used in the same way by various stakeholders. Many food terms have absorbed emotional potency by those who either want to proclaim their food identity, or those who are critical of food identity being an important personally defining characteristic. Most importantly, I must explain what I mean when I say "food." Food is life. All food begins with living beings and involves the transfer of life energy from those living beings to us humans.

We can't survive without food, without accepting the gift of energy and nutrients that are formulated and then offered to us by our food sources. We are sustained by food. To continue to thrive as a species, we must rediscover an appreciation for this life-giving energy transfer and its sources. Whole foods, foods that are in their original form or can still be recognized for their original form after minimal processing, make it easier to visualize and feel the transfer of life energy because the complex nutrients they contain have not been manipulated. Food is life in a tangible and consumable form.

Food sustainability then is to recognize, raise awareness, preserve, honor, promote, and celebrate living beings that are our food and nurture them so they continue to provide for us. It is restoring balance to a circle of energy transfer that has been happening since the first humans roamed this planet. Since that time, our species has developed and created increasingly complex societies, some that still maintain a close relationship to food sources, and others like the United States where much of what we eat has been processed to the point that we can't identify the food beings who contributed to it or how much of our meals are manmade additives and fillers.

What many people consider as food today is so far removed from the whole foods that provide the complex nutrients we need for physical, mental, and emotional health, that our country is plagued with several different and preventable disease crises.[1,2,3] Beyond that, millions of people are denied access to healthy whole foods due to established political and economic policies that overwhelmingly disadvantage based on race and class.[4,5,6] Disregard for the importance of understanding food is life and food is living causes us to approach food insecurity with a quick fix attitude. Those who can afford to throw imitation foods at the problem in waves when emergency support is called upon, but food insecurity remains and grows.[6,7,8] Maintaining and improving safety net programs doesn't solve the problem of food insecurity; a radical shift in food attitudes is needed.[4,6] To trigger that shift means thinking about food differently.

I haven't always recognized food is life. The transformation in perspective took time. For the first two decades of my life, food was invisible except for a handful of rare notable experiences that I will describe in later chapters. I ate enough to survive, but what I ate and the experiences surrounding it made no lasting impression on me. I grew up in a single parent family where there was never enough money to pay all the bills, but not at the level of poverty that many experience today. I remember collecting cans and bottles for the recycling money ($0.05 per can), in order to splurge on submarine sandwiches from the local deli for a "special" meal every once and a while.

For the most part, my diet was made up of inexpensive processed foods. Cold cereal, Pop Tarts, boxed breakfast pastries, oatmeal cream pies, pasta and store-bought sauce, and sodas were the norm. Occasionally, a wrinkled rotisserie chicken in a black plastic boat from the prepared meal section of the grocery store would provide a break in the pattern. In the summer when neighbors shared zucchini from their garden, my mother would make bread with it. That was a fresh and rare treat, and today, the smell of fresh zucchini bread makes my mouth water instantly. Overall though, meals simply were not a regular social gathering time; we sometimes ate together but not as a consistent practice and what we ate was un-notable.

Once I left home, I continued walking a tightrope of food insecurity with the only change being that I now understood that I was food insecure. I supported myself by working in the dining hall of the college I attended and then as a waitress in chain restaurants. As a first-generation college attendee in my family, I did so entirely through the use of school loans, scholarships, and working part time with no financial support from my parents. I wanted to go to college because I believed there was learning available there that would inspire me; my family did not strongly encourage me that it was an important experience; the only comments made were questioning how I would pay for it. In the end, the debt I had upon leaving school impacted me much more directly than the education I received for many years, although looking back I can see the value of my learning experience as a subtle influence throughout my entire adult life.

It is ironic to me that as naïve as I was about food as a young adult, I would spend the next couple of decades supporting myself by working in restaurants. Since I didn't have a regular interaction with fresh whole foods as a kid, the fact that the food prepared in the chain restaurants I worked at came primarily from cans and boxes and was simply assembled and heated seemed normal. I was serving people meals, but food was still invisible. I didn't think about food as special. When I developed a rash from the oil used in one establishment that made the skin on my hands crack, peel, and bleed, I didn't think deeply about what else it might be doing when ingested by customers, I just used lotions to treat it and wore gloves. I couldn't yet see the connection between food and an appreciation for life, in part because living was such a struggle that the present moment was only about survival. Like many, I clung to the hope of a future where comfort and appreciation for living existed, and I would find it eventually.

As I developed my service skills, I was presented with the opportunity to work for a family-owned restaurant that considered food a valuable commodity. Where the franchise chains were focused on one pageantry or another to draw people in, and meals were just the perfect regularly repeatable time to grab the customer's attention and develop habitual attendance, I began to realize the food itself could be the experience. Since I was food insecure, I would often buy meals with an employee discount because it was cheaper than groceries, but I would always get the most plain or bland things on the menu. French fries, mashed potatoes, macaroni and cheese, and pot pies became my regular diet items.

As time passed, I moved to more expensive restaurants because the higher priced the meals, the more I could make in tips, until I was finally making a livable wage (provided I budgeted properly for slow business and increased living expenses during the cold winter months). At the same time in my personal life, I was introduced to how other families ate and celebrated with food. I was welcomed into a family that had large meals for every holiday, which was completely stunning to me. The time of year or holiday dictated the menu, and everyone worked together, contributing and enjoying.

I could see that food was an important part of the dynamic within their family. Certain dishes were prepared with extra care because they were someone else's favorite, and the kitchen was the place to be, full of laughter and stories. Massive turkey dinners with countless sides for Thanksgiving, and meatballs, manicotti, stuffed shells, braciole with pine nuts, lasagna, and garlic bread all made an appearance at Christmas. Easter featured both ham and fish complimented with everyone's favorite side dishes. Beyond that, most evenings throughout the year had a mealtime, too. Dinner usually had a meat, a vegetable side, and salad. Food was absolutely central to the concept of family. Either food was cooked with care to nurture another, or gratitude was given to those who provided the meal.

At this point, I was forced to learn more about how food tasted. I began working in a place that would make one of each of the dinner specials nightly for the staff to share and taste prior to service so we could speak to customers about the food clearly and with confidence. An entire world of sauces was opened up to me, and fresh vegetables cooked to perfection. I had never eaten much fish before, except for perhaps shrimp at a seafood chain, or fish and chips at a pub. I tasted avocado for the first time in my thirties. Food was terrifying and exciting. I needed time for my pallet to adjust, and I was often ashamed by my lack of experience. Many times, I claimed I didn't like something simply because I had never tasted anything like it before and the experience was so foreign.

Soon though I was hooked. I was falling in love with food. Working in restaurants allowed me to buy dishes that I could eat part of and take part home stretching them over multiple mealtimes. Still, I didn't think of food as living yet, perhaps because I hadn't begun to interact with it through preparation. There is definitely something about harvesting or using fresh ingredients to create a meal that helps to demonstrate the transfer of life energy. In the last twenty plus years, I have had some amazing teachers who showed me proper knife skills and coached me through recipe testing, helping me build confidence and a love for drawing out the natural flavor in foods. I learned to use my senses to smell, hear, and feel freshness, and interpret that as valuable. Once I began to feel more comfortable with the mechanics of cooking, then the deeper understanding of food began to sink in.

For the talented chefs I was lucky enough to work around, fresh ingredients were always the key; dishes made from fresh whole foods explode with flavor that our body and brain naturally recognize as high in nutrition. As a result, I have eventually developed a deep appreciation for food (specifically fresh foods prepared with care) as a tool to pass on the life-giving energy that can make us whole and fill us with creativity and peace. I now strive to never take for granted a shared meal or the privilege of providing another food. That is my food story boiled down, all the various ingredients over the years contributing flavor to the stock, providing me with a view of food that became the base for this book.

While writing this text, I discovered that discussing food sustainability and security with others can be very challenging because of our diverse personal food biographies. The path to discovering food is life is not smooth and paved; there are ruts and prickers as well as beautiful vistas and moments of quiet perfection. The best, and worst, parts of expanding our appreciation for life giving foods are that each person's food journey is unique; there is no set guide. When we share our views with others we may be encouraged to dig deeper, or we may lose our passion based on a lackluster response. We ourselves may be the provider of feedback that stokes or smothers a food is life ember for another without realizing our influence. Familial or religious traditions contribute to one's base perceptions, and then life events apply filters that can either enhance or mute the flavors tasted. Our diverse backgrounds drive our values, which in turn impacts how we think about food.[9] As a result, we all come to the discussion of food sustainability and security carrying emotional weight regarding food.

A second stumbling block to investigating food issues is the human tendency to break complex topics down into bite size pieces. While it is important to comprehend the minute detail of a specific problem in order to deal with nuances, a solution formulated only on that narrow viewpoint isn't likely to stand up to scrutiny from another perspective. Beyond that, real solutions look at the root causes of food sustainability and security issues, not just options for treating the symptoms. While personal perspective and inability or unwillingness to consider the depth and breadth of the multiple problems associated with our foods systems make for a complicated dialog, the language and terminology we use are also enormous hurdles to surmount.

Perhaps the way we use words poses the most danger because of how they structure our perception of food. We have become so disconnected from our food sources in this country that even experts who promote a closer relationship with food often use terms such as commodity, resource, product, and *it*. Language helps us express our values even when we aren't thinking about them. Words are powerful, they can crush or cut deep; they can also inspire, stir emotion, and motivate. What we say builds upon itself; negativity breeds more of the same, and positive mantras blossom forth with hope and empowerment.[10,11] The way we speak about food demonstrates this dynamism; to transition from communicating about food as "it," inanimate and lifeless, to the complete opposite, requires a radical change in awareness and purposeful expression through speech. It has even been suggested that our current language may not possess the words to truly consider or acknowledge the capacity of plants to act with intent in partnership with us for survival.[12]

Food gives us life because it comes from living beings. If we embrace this sentiment, then it becomes so much easier to understand that every human has the same right to food. Food does not discriminate, the nutrition provided gives life energy to every human body, not only to some who can afford it or access it. It is the policies and practices of humans that limit access to food for other humans.[4,6,7] The way we use language to remove any trace of life from food and place it on the same level as many of the basic luxuries we consider necessities, allows for food to be weaponized. When we place more weight on oil, electricity, technology, or entertainment than we give to the food energy that keeps us alive to utilize those things, we demonstrate our priorities are completely upside down.

Reorienting ourselves to give food the proper respect and build food systems reflective of a partnership with other living beings necessary for our survival is crucial. This book looks at food from a variety of disciplines that we are all familiar with to show how they are interconnected and that it is time for change, and we can begin now in the small things we do daily. I believe that regardless of social or economic position, every individual can look at food a little differently and, as a result, make changes to how they interact with food in order to improve their experience of life, as well the lives of those around them. I don't think it's all about changing one's diet as the first step. Make no mistake, I think a diet consisting primarily of fresh whole foods with a leaning toward seasonal choices and respect for ancient cultural knowledge is optimal, and I hope many people see the value of that from the evidence I present throughout this book; however, the first step begins by simply becoming more aware. Thinking differently doesn't cost money, or even that much time; it does take some effort and self-reflection though.[13]

It will be helpful moving forward to look at and clarify the definitions for some commonly used terms related to food. Sustainability in relation to food simply means "the quality of not being harmful to the environment or depleting natural resources, and thereby supporting long-term ecological balance."[14] Food is life, it all comes from living beings; those beings need a healthy environment to thrive, and humans have the ability to help promote that healthy environment or destroy it. What I have found in countless conversations with people is that food sustainability has come to represent something larger and very diverse to different people, and that complicates things right from the start.

When each person understands sustainability differently and speaks from a place of interpretation they are comfortable with, the conversation becomes an exchange or debate over other food-related issues before establishing common ground of the basics. By allowing other food issues to be substituted in place of or discussed as part of the concept of sustainability, the word itself begins to take on an overarching and powerful momentum that represents a social movement to some and a misunderstood conglomeration of assertions that trigger defensive attitudes in others. For this text, the term has been tamed and put back in its place. Sustainability is about growing food (**and consuming it**) in a way that supports the environment naturally so food can continue to be grown (**and eaten**) in a healthy way continuously far into the future (**by everyone**). We all bring personal values to the table and perhaps feel they are not respected by others, but in the end, we are all at the table because we must eat to live. This truth grounds us to others despite radically different backgrounds, opinions, and concerns.

Perhaps the most baffling use of language to me is the food phrase "the Alternative Food Movement," which encapsulates any and all ideas meant to call attention to food as living and essential for survival. Fresh whole foods were the first foods; they have fed us for millennia, yet they are considered the "alternative" by our society now. If we reset our perspective to think about food as life, I think we can recognize how our language has helped to normalize highly processed food products as equal to or better than fresh whole foods despite the evidence overwhelmingly showing that these food products are negatively impacting our physical, mental, and emotional health.[1,2,3,15,16,17] Highly processed food products have given rise to epidemics such as heart disease, diabetes, overweight and obesity, and malnutrition. Our only option to combat these health issues on a national level is to return to whole fresh living foods as our mainstay and recognize the irony of calling them alternative.

Other terms such as organic food, food grown according to a certification process that only allows specific chemicals to be used on or near food sources along with methods that promote nutrient recycling,[18] are also to be considered in their purist meaning. Clean food is simply food grown free from chemicals, when organic certification is not achievable for a variety of reasons. Health means freedom from disease or ailment and with vigor and vitality.[19] Health is **not** a term endowed with power for use as condemnation or criticism of another. While some individuals have adopted these terms in their daily language in order to endorse their alignment with value sets that they see as positive, we must recognize that it has triggered an equal and opposite reaction for others. Likely not intended to belittle or disparage individuals who may be focused on concerns other than a positive food experience, defensiveness is a normal reaction to the perception of accusation or disapproval. This is reinforced by examples of the terms "health" or "target weight" actually being used purposely to criticize or castigate.[4,11]

The result of terms burdened or perceived to be imbued with emotional weight is discord. Instead of stepping back from possible misunderstandings in communication, in our culture, each side tends to dig in and accept their role as adversary. Those who see food as more than the most basic ingredient to survival, rather as a tool that can help facilitate a pathway to a fulfilling life experience and who want to share that revelation, feel justified and even obligated to speak out.[5,7,20,21] Their counterparts on the other hand can't allow themselves to relate or show vulnerability through curiosity about the power of food beyond basic sustenance. They must save face and maintain a defense against a threat that feels personally directed at them that they are "unhealthy" compared to their foe. In that survival instinct situation, they are significantly limited in the ability to think critically about their food choices.

To complicate further, infighting also exists between various parties who value a relationship with food, but harbor contention regarding the hows and whys of the paths taken to get to that shared concept.[4,7,22] The second-generation immigrant whose food culture is ingrained in their identity as part of their upbringing and the affluent young adult whose has recently discovered food, share a base value. Still, the conflict between them is real. As the first fights for the right to honor their culture, the second develops their ideas from a canvas tainted by a lifetime of advertising, sometimes bumping into understanding like a toddler learning to walk, and they will both likely offend the other at some point with their opinions about what food is "good." Food is the common ground though, and if we all take a step back and remember that, we can use respect and empathy to overcome obstacles in defining the qualities of food.

It is important to make connections at the personal level if we have any hope of instigating change on a larger scale. Our society's current view of food as a commodity results in both eaters and growers of food suffering economically and health-wise at the mercy of the agricultural complex that controls every step of conventional food procurement.[4,6,7,8,15,21,23] Because we all must eat, we end up contributing to the commercialism that is compromising the sustainability of our food systems. This is where a slew of additional emotionally charged terms come into play such as food security, food justice and food sovereignty.[4,5,6,23,24,25] Highly processed imitation foods are pushed as a solution to food insecurity by those who will profit from their consumption without consideration of the health impacts.[6,7,8] There is no foundation or central idea regarding enough fresh and culturally appropriate food for every individual as a base line.

Global organizations like the FAO (Food and Agriculture Organization)[3] and the WHO (World Health Organization)[1] state food is a basic human right, but that concept is not promoted consistently by the governments or citizens of all countries. Meanwhile, food insecure populations are represented by all manner of food organizations (not for profit and for profit), as well as academic research projects, and social support programs, with varied effectiveness. Food (and access to food) has become a weapon wielded to gain or maintain control of power and money,[4,5,6,7] and those who are wounded by the blade not only suffer personally but also become a combined weight on the systems that cut them in the first place. As time passes, the "us against them" mentality festers, and everyone gets assigned to a side even if they don't know there is war coming and the frontline battles have already begun.

I use the works of many popular food writers as citations in this book, and at the same time other resources I cite have openly criticized these writers as being blind to food justice and food sovereignty issues. In reality, all of these resources are trying to call attention to the weaknesses in our food systems, they simply have different perspectives. While raising awareness to the plight of the hungry is necessary, an individual first needs to be in touch with their own food mantra on a conscious level to take note of the concerns of another. I personally understand what it feels like to be food insecure and at the same time not have an appreciation for fresh healthy foods; there are many layers to the issues plaguing our food systems. I feel as frustrated as the activists that this problem has developed, persisted, and grown over the decades, rather than be addressed by individuals and community, state, and federal leaders.[6,7,8]

At the same time, I have friends who have never experienced food insecurity. They had fresh foods as children; they have a solid food culture passed down through family traditions. They may take some level of care in their food choices as a regular habit, or they may reject the privileges they had and choose to eat the imitation foods they were denied as children without ever thinking about the why behind their actions. The affluent individual can only look at food systems from a perspective of availability of food choices because food insecurity is such a foreign concept to them, and as a result, they will tend to resist engagement with the conversation in a meaningful way. It's like trying to explain to someone who has always lived in an urban environment and taken public transportation, the importance of a reliable vehicle for someone who lives in a rural location, or vice versa. Those two people have never walked a mile (or a city block) in the other's shoes, so they have no reference point with which to emotionally connect with the other.

What I hope to do here is show that no matter what approach one uses to examine food systems, it is important to first consider food from a living and life-giving perspective. From that common starting point, our individual curiosity profiles can be used to explore the various disciplines that impact food systems to better understand the complexity of food system fragility. For example, food security cannot be solved simply by growing more grains as a commodity product because that does not improve the issue of access to a complete nutritional diet for all.[6,7,8] Raising awareness about eating a diet of primarily low-calorie high-nutrient whole foods can only do so much if those foods aren't available to huge groups of people living in urban and rural food deserts.[4,6] Observing cultural or religious food rules depends on labeling regulations based on consumer wants and needs instead of complaints and demands by food processors.[8] Respecting Indigenous food culture and learning sustainability from people who practiced it for countless generations can and should be more than an educational exercise in schools.[13,25,26,27,28,29]

The simple truth is that what we are doing now isn't enough; millions of people are hungry, malnourished, sick, and often at the same time overweight or obese.[1,2,3] We need to better understand that food security is not only about the immediate need, even though that need is very serious. We are all in a sinking boat. Those on deck might not have wet feet yet so they don't understand the danger. Those below deck are legitimately focused only on trying not to drown. As a whole, we must work together. We cannot become so engrossed in trying to patch the holes in the sinking boat that we don't also pay attention to our course and make headway to a shore quickly. Everyone has a part to play in building a system that eliminates food insecurity and allows food sovereignty for all.

Without a cohesive base from which every individual can relate and come together, the countless food initiatives sponsored by localized groups can't demonstrate lasting and sustained improvement for others to model after, and they receive no support to stick with the successes achieved. Food system healing needs to be a mission everyone understands and supports in whatever financial and cultural metric they fit into because we ALL have to eat to survive, and what we eat matters. We must develop an overall measurement for progress and create accountability for ourselves and our communities. I believe that starts with thinking about food differently and recognizing that what gives us life does so because it contains living energy. Life energy is constantly being passed from one being to another, from species to species, and our responsibility in receiving it is maintaining gratitude and humility.[20,28,29,30]

From my conversations and years of research, I think the reasons some shrug off the importance of sustainable food systems is twofold. The definition of food itself has become blurred, and the complexities of nutritional impacts on the body are subtle. In large part due to advertising, whole fresh foods are often thought of as inferior to the food substitutes that industrial food processors create to capture our dollars.[9,15] The term "value added" used by industrial processors will be discussed later in more detail for its narrow interpretation of what is valuable, and that the nutrition lost during processing is considered irrelevant or superfluous. To some degree, we know we are being seduced, but the overall pervasiveness of marketing is unavoidable, so we tend to tune it out of our conscious thoughts and forget how much it influences our subconscious.[10,23]

Perhaps it is that same habit of dismissal that keeps us from noticing how our bodies react to what we ingest. A pang of hunger might be hard to ignore as a signal, but a foggy brain is rarely linked to what one ate for their last meal. The long-term effects of a diet poor in nutrients and high in substitutes are often heart disease and diabetes,[1,2,3] but millions of people suffering from these conditions aren't talking about the specific meals they ate and the clues they felt after that suggested a larger health problem might be developing. How we sleep at night, how our muscles feel in the morning, and our ability to focus on a task are all direct messages from our body in response to what we eat. Beyond what we eat, how we eat also impacts our ability to listen to the whisperings of our bodies. Eating meals in a rush, on the run and distracted by other activities smothers our senses. Alternatively eating a meal with few distractions and a focus of gratitude for the nutrition being given by the food sources and received in a thoughtful moment sets the stage for our bodies to sing a song of joy[20,21,22] or convey dysfunction in a critique of what has been consumed.

In the United States, we are both hypersensitive to advertising that promotes the celebrity nutrient of the week and dismissive of pain and discomfort as messages from our bodies related to what we eat, unless the problem becomes so bad we can't function. Changing how we think about food both when choosing our meals and dedicating time to self-care when eating can help to nudge us toward recognizing the body's reactions to the food we give it.[11] Practicing active awareness whenever we ingest something can also help trigger empathy for those who may not have what we have. One of the most basic traditions at many family celebrations is to give thanks for the blessings to be enjoyed. Extending this to a conscious thought of gratitude for our food source whenever it is consumed is an easy shift anyone can make, and the time it takes up in one's day is miniscule. Once that becomes a habit, then thinking about food differently can really begin and choices can be refined to ensure our diet meets all our needs, not just our hunger.

This shift in awareness can further help us learn that food insecurity is not exclusive to those least financially secure within our population. Food security is not just a matter of having money available or locations within our neighborhoods at which to spend our food dollars. Food security is impacted by weather; farms on the west coast face increasing wildfire risk, drought is impacting a larger portion of the mid-west annually, the strength and frequency of hurricanes jeopardize the viability of farms in the south and all along the east coast, and flooding can cut communities off from food deliveries in any corner of the country. Weather emergencies can impact everyone regardless of financial status.[31]

Beyond weather, a new food system risk that impacts all of us has recently unveiled itself due to the global COVID-19 pandemic in the form of supply chain disruptions. Fresh vegetables can become even harder to find, with price increases that reflect their scarcity, widening the gap further between those who have access to a nutritionally healthy diet and those who don't. Realistically, the fragility of current food systems put both individuals who struggle in the present to access fresh healthy foods, and those whose affluence may induce a false sense of security with regards to food access, at risk. Strengthening our food systems is everyone's problem, and we can all do different things to help build a stronger tomorrow.

Ultimately, consumers must be the driving force through participation in those solutions as companies focused on profit will never be the leaders in addressing food security.[7,8] The actions of every single individual living on the planet works either toward or against improving food security, whether they realize it or not. A passive stance supported by the false sense that food is always available at the grocery store or restaurant is a precarious position to be in. The good news is that we are not starting from scratch. Across the country in many places where food insecurity has been a present reality for some time, small groups have been using innovative methods to grow fresh whole foods that promote a healthy life experience.[32,33,34,35]

Addressing food system issues involves a great many factors. Defining food security and sustainability is both simple and exquisitely complicated. Every individual has their own perspective, culture, and financial status that impacts how food security and sustainability applies to them personally and that must be reconciled with the larger community as a whole if solutions for fragile food systems are to be effective. We must agree that everyone has the right to fresh healthy culturally appropriate food. Continuing to learn, think critically, and remain focused on the problems to be solved takes time, but it is necessary in the present moment. Stepping back to ensure that all angles of the problem are considered is perhaps most important. An interdisciplinary approach allows us to evaluate a variety of food system issues in relation to one another and find solutions for lasting improvement.

An interdisciplinary approach, considering the complex food system using a combination of viewpoints including health and nutrition, science and technology, political and economic policies, religious and spiritual beliefs, and cultural and personal values, helps identify benefits and drawbacks to every food partnership opportunity. It offers a comprehensive foundation to evaluate a "sustainable" hydroponic garden as a technological resource during winter months in colder climates to provide fresh produce. A hydroponic garden can reduce some of the petroleum resources used and waste generated by uneaten product compared to produce shipped in from other regions or countries. This is balanced by a shortened growing process that is unnatural and uses manmade chemicals to fertilize, and a final product that is less nutritious than that grown in the ground even though it is fresh.[36,37] When we look at it this way, we can fit hydroponics in as one supporting part of a larger overall plan to improve access to fresh whole foods, but not the complete solution. There are good and bad aspects to almost any idea to improve our food systems, but with individuals thinking critically from an interdisciplinary perspective we can reduce the negatives and increase the positives finding a healthy balance.

As I have said, the first step is simply to think about food differently. Every time you put something into your mouth to eat, think about what it is. Is it a fresh whole food or made from a recognizable fresh whole food? Is it a processed food? Why did you pick that food specifically: taste, preparation time or process, convenience, hunger, boredom? How do you feel while eating: happy, satiated, tired, stressed out, guilty? Before any changes should be made, one must find their baseline and decide why they want to improve their food experience. Eating differently for the wrong reasons because society or culture made you think you should (dieting) will never succeed. Building a stronger sustainable food system means caring about your relationship with food now and in the future. Once a benchmark is established, it's easy to make small but meaningful lasting changes that begin to build on one another.

Here is an example from my own experience. For many years, I avoided meat in my diet primarily eating a lot of fish. I did not think of myself as a vegetarian so much as I didn't really prefer meat. My partner prefers to eat meat more often, and as we share meals, I am faced with eating meat more than I normally would. While researching this book, I read a great deal about industrial farms that produce meat products: the high levels of cortisol (the stress hormone commonly associated with heart disease) in industrially farmed meat, the unnatural diets fed to the animals that makes them unhealthy and which ultimately we consume through the meat, the way that these meat factories use specific terms to disassociate the final product for consumption from the animal sacrificed to produce it (reducing the consumer's ability to connect and feel gratitude), and how government subsidies help support these producers by keeping prices artificially low while the climate impacts become increasingly costly. All of these factors supported my disinterest in eating meat available in the grocery store.

Instead, I began looking into local farms that raise their animals in a natural environment, feed them natural diets, and humanly harvest the meat in the least stressful way to minimize the cortisol and adrenaline that end up in the meat. This type of meat is often called traceable because you actively search out all the details that determine its healthiness. My friends joke that I like to know my meal's name but realizing that what I am eating came from another living being, working to confirm the living being had the best life possible, and ensuring I pay the farmers who do the extra work to provide that clean meat to me is important. I can honor the animal as my food source by ensuring I eat it thoughtfully and with gratitude. My partner and I also reduced our overall weekly meat intake so that at least half of our meals are vegetarian, and we eat smaller portion sizes supplementing meals with fresh vegetables that helps reduce our impact on the climate, as meat production is a considerable contributor to greenhouse gases and water contamination.

While our choice to do this might not have a measurable impact nationally or globally, consider how many people we might influence either by providing a vegetarian meal when we eat together or simply raising the awareness of others who might never have considered what buying packages of industrially produced meat from the grocery store really meant for their health or the health of the environment. Patience is important when planting a seed as any gardener can attest to. Sometimes, the sprout will pop up within days, but other seeds take longer to germinate and need to soak up the sun and get regular water before they present their delicate head for all to see. Even then it is important to protect the tender seedling from getting trampled or nibbled so that it has a chance to grow and flower.

In the following chapters, I approach the concept of food is life from various disciplines to show that understanding this concept is the key to rebuilding our food systems and ensuring everyone has the opportunity to be fed and nurtured by life giving foods. In fact, the perspective that food is life giving is strengthened when approached using a combination of disciplines that highlight the magnificent complexities of cyclical life energy exchange between all the living beings included in our food systems. Health and happiness are within reach and a diet of fresh whole foods shared with those we care about is the path to get there; feeding our bodies well also feeds our souls.

CHAPTER 2

VALUING FOOD

Every individual in our global community has a unique set of values by which they structure their life. Social groups often come together due to shared values, children are taught views held by elders, and life decisions are made based on what one considers most important to them. Conflicts between and within communities find their roots in disagreements over basic values. Our values impact every single aspect of our lives, whether we are paying attention to them or not. With regards to food, it may seem logical to say that each individual values food. After all, none of us can live for extended periods without it. In reality though, a dependance on food for survival does not automatically translate to an attitude of respect, gratitude, or valuation. An argumentative person might even say that the body can be kept alive through artificial means being fed intravenously, but those situations are almost always accompanied by a lack of animation and enjoyment day to day. Therefore, for this purpose, they are not living so much as existing on the fringe which is different.

If food is considered a universal value, along with water and shelter, rationally, all the other comforts we surround ourselves with should complement the value we place on food. Yet when we examine the local, regional, governmental, and global communities we are a part of, we find that food is generally ignored, undervalued, and even weaponized.[4,6,9] Food substitutes have become so prevalent that in the United States, many people do not even consider the difference between counterfeit and real food when picking their meals.[38] Instead of seeing food as life giving, it is viewed as time consuming and consequently devalued.[39]

There are pockets of people who have become sensitized to this transformation both intellectually and out of physical necessity that are trying to regain a healthier perspective with regards to food.[40] Locavores, pescetarians, vegetarians, vegans, home gardeners, CSA (community supported agriculture) members; the list goes on. The labels they claim and the rules they create to define their values separate them from the mainstream and highlight the lack of values held by the majority. At the same time, the dominant populations in some industrialized countries led by the United States regularly mock, disparage, and try to "aid" Indigenous groups who demonstrate the strongest alignment with valuing food.[36,41,42] How did this divide regarding the way different groups value food become such a massively deep fracture within our global society? We all still need food for our brains and bodies to function as one unit; we have not developed the cellular ability to transform sunlight into caloric energy like plants do.[43]

Let's start from the perspective of those most connected to their food sources to try and identify if and where the valuing attitude for food was lost or has faded. Although their numbers are shrinking to extinction, Indigenous hunter and gathering groups from all over the world are intimately tied to the natural surroundings they live in.[28,29,30] Their nourishment comes from the plants and animals they harvest and eat; they take what they need resulting in a sustainable food system, and they are thankful for the resources that the Earth has provided to them. They value their food because the core activities of their day-to-day life all revolve around some aspect of acquiring and enjoying food. Money is not as important as food because money can't be eaten. The lands they live on are being taken from them regularly in the name of progress; otherwise, their lifestyle would be able to continue indefinitely because they take only what they need to survive, acting in concert with the environment around them. In areas with a little more exposure to development, Indigenous subsistence farmers maintain a close relationship with the Earth and also live within the limitations of the natural resources around them. They participate in the exchange of food for money as a means of survival, not profit.[28,30,36] They are a part of their local food system, and they take appropriate action to live sustainably.

From there, the trend continues; developed countries with strong cultural ties related to food support local farmers trying to balance sustainable farming with profit.[6,9]

The more industrialized the community, the more value is placed on money, and the less attention is paid to quality grown foods and sustainable farming practices. Government programs in the United States that pay farmers not to grow food in some fields and provide additional pay to grow only mass quantities of a few grain products in others, might have once seemed the most extreme example of valuing currency over nutrition. Today, a whole new level of the dollar's strength to erase value systems has begun to emerge as GMOs have been introduced into the picture.[44] For the average consumer, the array of substitute foods has exploded in the market, and it is hard to recognize real food at all.

Healthy and nutritional are buzz words used to market highly processed items that bear no resemblance to fresh food. If our cultural knowledgebase does not draw a line between edible items that humans create based on our current understanding of nutrition requirements versus items grown by other living beings with complex chemical structures and components that work in concert to provide nutrition, then how are we as individuals to delineate between them? Further, organically grown whole foods possess qualities beyond chemical compounds in my opinion, transferring something less tangible—the very essence of life. While some may recognize organically grown food as superior and believe that concept to be obvious, buying patterns suggest that for the majority, money is overwhelmingly considered more important than nutrition.

Often, the cost of organically grown real food is more expensive than their counterparts, suggesting that value and price are related, but in truth, it's more likely that the market sees choosy consumers as a niche they can exploit instead of a cultural acknowledgement of quality. For most consumers the reality is that our mental and emotional connection with food has become severed.[21] The leading societal attitude of food not being worth the cost is evidence of our priorities.[45] Because food provides the nutrients needed to live, that should imply that society should reject money over food. Instead, we have embraced substitutes that fit into a cost spectrum that we find acceptable and rejected real food. The value of food has been lost.

I have often heard the statement that there have been no repercussions as a result of the "substitute food - profit over sustainability" model in my research of this subject, and I can only scratch my head and wonder if the effects are even more serious than I understand at this moment. As I have approached this topic from varying viewpoints, devaluing food is possibly the most critical mistake the human race has ever made. Any other miss-step can be traced back to food, as I will try and demonstrate in this text. Economically, scientifically, politically, culturally, spiritually, physically, and mentally, the value we recognize for the food we ingest has a direct impact on our lives.[46,47] Each of these disciplines will be covered in greater detail as this book continues, but each is also linked to our individual established value sets. The decisions we make about when and what to eat reflect how we value our own bodies. Our attitudes and actions surrounding mealtimes demonstrate clearly where our priorities are focused. Our obsession with limiting time spent on feeding ourselves compared to the time we are willing to commit to scrolling and clicking electronic devices is an indictment of our value shift.

We tend to limit academic consideration of sustainable food topics to only one discipline at a time, forcing us to ignore important relationships and consequences between them. This habit could be a result of the established reductionist methods that scientific study has evolved into within the last two centuries.[48] Society's values are widely influenced by science as well as cultural and religious conventions. Our inability to consider the complexity of food sustainability may also be a physiological limitation derived from lack of balanced nutrition because of consuming primarily substitute food sources for much of our lives. This idea hasn't been studied in depth yet, possibly because we still haven't come to terms with the other more obvious impacts of rejecting real food in favor of substitutes, but the combination of all the sources noted in this book demonstrates that the more we learn the less we seem to understand about how important food is. Health issues continue to emerge, and new cancers develop faster than older types can be cured throughout industrialized countries.[37,49] Most scientific research is also funded by those poised to make the most from food substitutes; therefore, studying the negative impacts of a lifetime spent eating food substitutes isn't likely to get enough financial support to even begin collecting data.[6,8]

A healthy balanced diet is known to both prevent and heal, yet the government's recommendations for a healthy diet are influenced by political lobbying rather than medical evidence.[8,21,50] Our ability to evaluate what is best for ourselves is foggy and blurred to the point that we simply follow the trends. Our society's diet is manipulated by the government, industrial farming conglomerates, and profiteers.[4,6] Although there is a minority at the top of this triumvirate of power players, they are not exclusively to blame.[51] Each time one of us reinforces the value set they have established, we strengthen their power. Each time we choose a substitute over a real food item, we place value in money and power over health, connection, and happiness.

Does that sound like a stretch? I will admit I didn't start this project believing so deeply in the importance of our connection with our food. I love fresh whole foods, cooked expertly by artists who draw out the most delicate and amazing flavors. I am passionate about the shear enjoyment of a dish so much so that the rest of the world melts away as all of my senses react simultaneously in the experience of the momentary magic in a bite. From the reactions of my friends, my ability to connect so intensely to food in that sense is pretty rare. It was partially their lack of understanding about how much pleasure food can bring that nudged me on this journey. My desire to better understand how our food systems work, and to discover a way to expand my own enjoyment of food further, combined with wanting to reintroduce those around me to food as I saw it, propelled me forward. I had no idea how terrifying it would be to discover that even those who have an interest in healthy eating might be blind to the bigger picture impacting our species. I never conceived that human beings might actually not realize food is a basic necessity, deserving of our admiration and gratitude. What I have discovered is frightening.

Please do not infer that I somehow have figured out all the answers or have discovered a secret that will bring eternal bliss. As I study food through various lenses, the questions continue to stack up. The most consistent question is "Why?" Why don't we embrace the solutions that we know will move us in the right direction? Why do we choose options that do not support our continued survival? Why are our values in so many areas of our society eroding? Why weren't those values strong enough to survive and drive our society on another path? Why couldn't we pause and consider the bigger picture at so many different historical moments that led to here. Why does nature keep providing us food when we abuse the ecosystems we are a linked part of? I don't know the answers, but I can see crossovers from one discipline to another. I think the value placed on food by society corresponds directly with humankind's disconnect on each of these other levels. I have ideas that some may consider provocative and eccentric, but that reaction is part of this topic as well. We must eat food to live active lives, so any action or thought regarding food is linked to our survival.

If we don't value food, then how do we perceive it? Perhaps the question answers itself. *It*. We perceive food as a *thing*. We are so far removed from the sources of our food that we don't recognize that all of our food sources are living beings. Even those who practice geophagia, the eating of certain soils for the mineral content, are ingesting the microorganisms living in the soil. Mother Nature, the community of non-human living beings that inhabit the planet, cares for us by turning sunlight and water into chemical compounds we can ingest as a life-sustaining energy source. The food we need to live was all living tissue at some point. Does this mean that to live according to a belief system in which we cause no harm to another, we must starve and die? The ecosystems of our planet evolved in a way that most living beings feed on other living beings in mutually beneficial relationships.

The difference between our culture today that includes groups who feel the need to restrict which living beings one is willing to ingest, and the system that developed naturally, is the balance and respect with which they participate in the system they are a part of. Ecosystem harmony prompts that when food resources are low, birthrates slow and populations shrink and when food resources are abundant, the rest of the system benefits. In a sustainable system, not one species acts outside of the balance of the whole. Conversely, in our current unsustainable situation, we have created substitute foods full of empty calories and lacking in proper nutrition that trick the population with a sense of abundance.[6,21,37] Consequently, our population has exploded while the environment in which our food grows is withering. We are over taxing and undercutting our own survival.

Does the idea of valuing food as living beings seem scary? If so, why? There are religious taboos that limit food choices based on classification of life form that may trigger a pause,[52] but how does one reconcile eating some life forms instead of others? If each bite of food is a gift from another living being, does that cause pangs of guilt? To make the leap and consider oneself part of a group of diverse living beings who provide for one another in a network, exposes another fundamental flaw in the human identity. If humans are indebted to other life forms placing us on a level playing field, then the perversion within our species to consider other humans as less than equal because of a physical or economic attribute becomes transparently ridiculous.

If we can muster enough humility to grant value and stature to a plant that produces delicious and life-sustaining nutrients, then how can we possibly look at another human sharing the same need for food as somehow fundamentally different or less than ourselves? The talents, strengths, and weaknesses that make each human being unique compared to another is a mirror of how nature uses redundancy by layering multiple species to perform or accomplish specific functions. Human individuality provides the opportunity for us to approach situations from many angles and develop a wide range of solutions to issues like nature's preference for redundancy. We should embrace our individuality for its merit as an evolutionary asset while honoring our unity within the natural world as a species. Categorizing and rating one another based on our differences rather than recognizing that every human is human, and therefore is an equally valuable member of our species is a waste of time and ultimately limits our evolutionary potential. Maybe this is a clue to why we are encouraged to devalue food in our current culture. The danger that it might give birth to a more equal society could be truly scary for some.

Here is a moment when we can pause and see the strings tugging at us from the past; the certainties we base our opinions on. Humans can't be equal to other life forms; we are so much more advanced intellectually. We have emotions and a sense of self; we are sentient and other living beings aren't. Where does this mentality come from? Some religious teachings, like that of Christianity, suggest that humans have an elevated status over the rest of nature. Our language systems reinforce this concept far beyond the reach of religion.[12] Houle, a Professor of Philosophy in Canada, sums up how language influences us succinctly, "Through...linguistic choices, no matter how banal or how quickly they are uttered, our own ways of being in the world shift. There is activating and deactivating of our impressively varied capacities to be: with, or apart from, observer of or experiencer in, questioner of or teacher to, receptive to or hostile towards, loving or blocked, from the total situation in which we find ourselves."[12]

Conversely, a variety of other belief systems are structured on a model where humans are part of the many species functioning together for the benefit of all.[13,20,28,29] Their messages are a chorus united in the belief that we must respect and protect the Earth as our Mother because we are dependent on her for survival. These belief systems are built on the notion of humans recognizing the value of the life sustaining resources provided, and the responsibility we have as part of the overall system; in short, dependence and equality, yet our language restricts the adoption of these values into the mainstream.[12]

If contemplating this perspective is discomforting or un-nerving, pause and consider why. Do you feel as if you are losing something by considering the equality of various species? Is your concept of your own worth and value as an entity in danger if you consider food as living beings providing for you? Based on the endless conversations I have had with people on this topic I think this type of emotional response is common. Regardless of education, age, ethnic, and cultural backgrounds, this discomfort is universal. The human race has evolved to this point with a self-awareness based on supremacy. This seems ridiculous to me given that we are so delicate and fragile. We can't generate our own nutrition within our bodies, and when we try to duplicate and manipulate what the living beings around us provide for food, we develop subtle health issues that compound into the crises we see today.[4,6]

I keep returning to the same query. Does recognizing the value of other living beings have to take away from our own value? If the current perception of ourselves is based on a mutated notion established in religion, and it can be shaken so easily by the simple suggestion that another life form may be our equal, we aren't very stable. In fact, the frailty of this mentality is astonishing. After much consideration, I believe that we would actually be emotionally stronger if we embraced our position within the ecosystems we are a part of. We evolved within these ecosystems. There was a gap we as a species filled, a need we satisfied. As part of that system, we have legitimate value. By recognizing the value of other life forms we partner with, we increase our own value. Visualizing ourselves as separate and dominant creates a situation in which we are segregated from the functioning system, and therefore at constant risk of being excluded and becoming obsolete. This thought process has extremely strong parallels with how we treat one another within our species. That could be part of why we have such a strong emotional reaction to the concept. The simple truth is that being an active part of a functioning food system benefits us and validates us.[47,53,54]

Progressing on the premise that we begin to recognize the difference between real food and the pervasive substitutions monopolizing the market, and that we have decided to make a concerted effort to value real food, we are ready for the next step. In the United States, the overall population is so large and divided by myriad geographic, governmental, and cultural categorizations, that how each community instigates change will be different. Early examples of those who are testing new societal configurations born from a commitment to real food can take on a variety of appearances. The University of California Santa Barbara mission,[55] the North Carolina grocery project,[56] and the Illinois local food study[57] approach the improvement of their food systems from radically different angles, yet they share one attribute. Each project is led by individuals who value food. The roadblocks and struggles identified in each situation could only be overcome using an open-minded approach that rejected the established norms. Changing existing systems is not easy; it takes time and a pledge to not back down, rather to push through with ingenuity and dedication.

Do we need to go back to a time when each family unit produced all their own food? That would be impossible.[21,23,36,44] The human species' global population has grown beyond the point where there is enough land for each small group to produce all their own food in a sustainable way. Alternatively, local communities can build a network that functions in much the same way, allowing for each individual to contribute without dedicating the extensive hours one would have to on their own.[32] Individuals who make the personal decision to value food must develop a support network. Ecosystems and the natural sustainable food systems that once existed within them, have overlapping and redundant connections that form a supportive web. That model should guide us well as we develop a new practice around choosing and acquiring foods. Working with various partners to ensure each has what they need, including non-human partners, will strengthen the entire community. Recognizing the value of the living beings providing our food triggers a chain reaction that improves our self-esteem, our perception of other human beings as valuable, and works to improve the environment we live in.[31] There is no downside to recognizing the value in food. In contrast, our current practice of devaluing food has distinct disadvantages including food insecurity for all.

Does valuing food mean spending more? This question doesn't have a straightforward answer of yes or no for several reasons. The question itself reinforces a link between money and food rather than allowing us to consider food as a relationship with other living beings. Because we can't acquire the nutrition we need by eating money, the relationship we have with food should be viewed as more important and different than our relationship with money. That may be easier said than done for those who have very little of either, but in those cases, the focus on food is even more important for survival. Personally, I have had to make the choice between buying food and paying bills and the very humbling experience of borrowing money just to buy a few groceries in the past. I have firsthand experience about how humiliating it can feel to work two jobs and still not be able to make ends meet.

At that point in my life, I had not even begun to consider that money and food should not be fused together, or that allowing that coupling was actually part of what was crippling me. I didn't have a clue that the substitute foods I was eating were dulling my thought processes and keeping me from finding a better path forward. I was fighting to find balance while systematically sabotaging myself with my diet. I was fortunate in that I wasn't significantly overweight and dealing with physical illnesses as a result, but I was battling the mental and emotional effects of empty calories. These invisible symptoms are no less deadly; they kill our spirit and our hope, and they trigger an apathy that holds the individual hostage. My experiences were in large part what has driven me to study how our society currently values food, and what influence it has on our overall enjoyment of our lives. As I learned to value food in my life, it became more and more clear how my relationships and interactions with others changed for the better. Confidence and empathy are nurtured with a healthy diet, forming the framework for connections.

Changing what we eat by reducing substitutes and increasing fresh food sources improves health and reduces the costs related to treating health issues.[37,53,57] The shift in money spent on medical professionals and pharmaceuticals to treat illnesses, as well as what is spent on gimmicks, fads, and diet plans, to money spent on real food purchased in season will result in more cash in everyone's pockets. The food may cost a little more initially, but the money saved by reducing what is spent to repair and improve our failing health will greatly outweigh the increase in food costs for most. As local food systems grow, the cost of food will be reduced because associated fertilizer, pesticide, shipping, and refrigeration costs will be eliminated.[57] There are many ways to access fresh healthy foods on a budget by shopping locally and eating seasonally; later chapters will cover this in more detail.

In my experience, viewing food as a partnership with other living beings compounds the benefits of balanced nutrition. Perceiving food sources as partners breeds feelings of gratitude. It is hard to describe the sense of comfort that is gained by developing a connection with the natural environment that nurtures us. Imagine the camaraderie within close knit family or social groups, except that this relationship is reinforced with every mouthful, during every meal or snack, every day. It is exponentially more intimate both in the frequency of contact and in contact type. With friends and family, we must rely on one another to be truthful and honest. When that trust is violated, it can cause deep emotional wounds. Similarly, what we ingest multiple times a day, good and bad, literally becomes part of us. Our bodies absorb the nutrients and/or toxins and react accordingly. The difference is we cannot distance ourselves from an afternoon apple or a substitute food picked up at a convenience store like we can a best friend who lied about something crucially important to us, or a spouse who ignores their vows.

If valuing food is so important, and most of the developed world has forgotten how critical it is to honor this relationship, how do we introduce ourselves to food again? That journey will be different for each individual depending on their current relationship to food. The conscientious eater, who strives to maintain a healthy diet already, simply needs to perceive their food and the sources they get their food from a little differently. Someone who claims they would eat healthier if it were easier may need to evaluate what it is they find difficult about giving themselves the best care possible, and acknowledge that this should be prioritized above distractions that may be holding them back. It may seem that the eater who doesn't even desire a healthier eating lifestyle would have the hardest time making a change, but this population's reaction to a new food attitude would be the most noticeable in the shortest time frame and help to reinforce the benefit. On a daily basis, I hear people around me complain of all kinds of minor ailments and general pessimism that I believe would melt away with a new food mentality and practice.

Making a change requires an effort, though it doesn't mean throwing every item in the cupboards and fridge out and starting fresh tomorrow. Compare the experience to meeting new people and dating. The first step would be to find a few social spaces where others with similar interests spend time (finding some whole fresh foods that are easy to prepare). The next step would be talking to various people (trying new fresh whole foods regularly). From there, connections will develop, and future meetings can be planned to learn more about each other (making time for a specific number of meals weekly that are made from fresh whole foods). During those "dates," some enjoyable activities play the backdrop while questions are asked and answered, checking to see if enough common values are shared to form the foundation of a relationship (sharing meals with friends and family and concentrating on receiving wholesome and pure nutrition and life energy from your food). This process will likely be repeated for weeks or months during which time trust and affection are developed. Finally, both parties are embedded in a mutually beneficial and reciprocal partnership. A perk of getting to know food in this way is that no one is waiting to see who will text first after each date. The process with becoming closer to food, however, is unlike a relationship that may grow stagnant and break down. We simply can't stop eating, and the potential interactions to get to know food happen many times a day versus once or twice a week.

Valuing food differently begins first with exposing one's self to fresh healthy food sources. Try visiting a farmers' market or farm stand weekly and talk to people there about why they are passionate about food. Start by simply asking what item the farmer recommends that week and how they would prepare it. Commit to trying new foods and try various preparations before immediately banishing something from your diet for eternity. An important key to remember is that substitute foods are developed and tested extensively and often loaded with sugar to induce cravings.[50] Fresh foods will taste different, and the body may need a moment to recognize and adjust to the new additions.[21,39] Fresh healthy foods have spent millions of years perfecting their gifts for us, and our bodies will react positively to them once the man-made substitutes become less frequent and distracting.[21,28] We have also been trained to expect a full or stuffed feeling with substitute foods as a result of the empty calories. The body keeps asking for more because it isn't getting what it truly needs,[37,50] and the speed we eat with doesn't allow us to listen to our bodies.[23] Start with foods that are easy to prepare and provide a feeling of satisfaction.

Most importantly, take note of what it is you are eating. Is it a leaf, a stalk, a root, a seed, or the flesh used to spread seeds? Research what the entire plant looks like. Offer thanks to the plant, the farmer who cared for it, the rainwater that bathed it regularly, and the sunshine that warmed it. The few moments it takes to give this consideration will affect the taste. At the end of this chapter is an "Introduction to Valuing Food Exercise" I used during a presentation at a wellness weekend. Many of the participants noted after that they had never experienced such an amazing amount of flavor in the simple act of eating an apple before. This candid exploration to reorient our perspective, opening up our potential for enjoyment is the first step to learning how to value food.

The act of eating influences every bodily function we have from thinking thoughts and learning, to movement and health. Exploring a new mindset with regards to our approach to food satisfies both a selfish need for enjoyment and a selfless approach to other living beings. We can be fed in every sense of the word and honor those that provide sustenance simultaneously with a mindful approach. Awareness, triggered by a mindful moment, awakens us to the incredible potential of being connected on the deepest level with our surroundings through an attitude of valuing.

Simply being mindful when eating can help to uncover which foods are fresh and designed to support a healthy body, and which are substitutions with little to contribute. The process of developing a mindful practice may feel uncomfortable initially. It is a completely a new way of perceiving ourselves as well as the food we eat. Learning to value food in turn causes us to value ourselves as the recipients of a level of generosity that is rare between humans, on a consistent daily basis. Although many examples can be found demonstrating the immeasurable depth of compassion that humans can express, consider that our food sources give of themselves every day. We have the responsibility to respect, admire, and protect them. Humankind has been part of this partnership as we joined the rest of the species already making a life here on Earth.

As evidenced by populations in industrialized countries, most specifically the United States, living up to the duty of valuing our food has been pushed aside. Initially, there seemed to be no repercussion, and we were encouraged to keep testing the boundaries. Now, we find ourselves discovering how a poor diet impacts every aspect of our lives from physical and mental health issues to low energy and satisfaction. Greed, profit, and convenience are promoted as values instead of gratitude and balance. Plants and animals that provide us food are abused and tormented,[58] rather than cared for and honored. I whole heartedly believe that we haven't even begun to understand the impact of our actions and the way our bodies operate. I believe that changing our mindset to value our food sources will provide plenty of proof regarding the benefits. If I am wrong and valuing our food doesn't create a universal mental and physical improvement within our society, there are no negatives to trying, and it might help us enjoy the meals we eat every day in the meantime. Humans benefit either way.

INTRODUCTION TO VALUING FOOD EXERCISE

This experiment can be done with any fresh fruit and takes just a small bit of preparation in advance and about fifteen to twenty minutes of your time. This exercise was expanded and adapted from a mindfulness practice described by Buddhist teacher and author Thich Nhat Hanh.[52]

Preparation: Purchase a piece of fruit that is as close to "in season" as possible in your area. *For my example, I use an apple because they are usually easy to find fresh with relatively local sources for most of the year.* Try to find something locally grown. If there are no local options, pick something that comes from your region or country, but be careful to choose it for freshness. Splurge on an organic specimen if possible. *The farther a fruit or vegetable travels to get to you, the more has been done to it to preserve the fresh appearance; these techniques can reduce taste.*[59]

In a quiet place, read each line of this exercise, pausing to really consider what is said. This can be done alone or in a small group with one person reading the instructions aloud. Don't rush.

- ❖ Look at your apple . . .

- ❖ Think about the seeds inside. Dark. Teardrop shaped. Soft and white on the inside.

- ❖ A seed just like that found itself in the perfect combination of moist soil with warming sunlight some ten to fifty years ago.

- ❖ That seed sprouted into a little stick poking out of the ground.

- ❖ The small tree likely less than a foot or two tall managed to avoid getting nibbled by animals while it soaked up sunlight, rain, and minerals from the soil for a few years.

- ❖ Its roots grew beneath the ground just as its trunk and branches grew above.

- ❖ Humans passing by could only perceive half of the incredible growth this little tree was involved in.

- ❖ Each year in late winter and early spring, the snow that insulated the roots from bitter cold also trickled refreshing moisture down so that the roots could manufacture a slurry of sugary sap.

- ❖ That sap is the life blood of the tree.

- ❖ As the weather warmed each spring, the roots shifted into overdrive pushing sap up through the living layer of cells between the bark and the wood of the tree trunk.

- ❖ For ten years, this process repeated itself and the tree grew taller, the trunk thicker.

- ❖ Growing in the spring and summer.

- ❖ Withdrawing into dormancy in the fall and winter.

- ❖ Last spring, within the dead looking branches, buds for the new season's branchlets and leaves started to swell.

- ❖ As the sap delivered sugars and water to the branches, the tree exploded with thousands of blossoms seemingly overnight.

- ❖ Tiny flowers the size of a nickel, each with five delicate light pink petals.

- ❖ Mustard colored pistils and stamens in the center of the flower sent a fragrant call out to neighbors in the natural community offering a tasty nutritional treat in exchange for a little help.

- ❖ Answering the call, bees, months, butterflies, beetles, and ants all came running and flying to the tree.

- ❖ They happily rubbed themselves all over with pollen as they ate their fill.

- ❖ The pollen was full of sugary energy-supplying nutrients that they need in order to mate.

- ❖ The insect community only ate what they needed, and they visited many flowers taking only a little from each.

- ❖ In the process, they also moved some of the pollen from one flower to another, and to other blossoming apple trees in the area.

- ❖ By sharing pollen between different trees, the seeds produced will have a combination of traits from multiple trees, making its offspring stronger.

- ❖ It is like having curly hair from your mother's family and height from your father's family.

- ❖ These insects are genetic assistants to the apple tree in nature, mixing up the gene pool.

- ❖ After the flowers are pollinated and the petals rain to the ground, the leaves unfold.

- ❖ Take a look at the bottom of your apple.

❖ Do you see the little puckered area?

❖ That is what is left of the flower.

❖ On the other end of the apple, you can see the stem where the flower bud was attached to the tree.

❖ Between the stem of the branch and the center of the spent bud, a little hard knob started to form.

❖ From April to September, for five months, the tree's leaves absorbed the sun's energy.

❖ The tree's roots soaked up rain when it fell softly enough to trickle into the ground rather than run off.

❖ Nutrients from the soil were drawn into the roots along with the water molecules.

❖ Those nutrients were sent up the living cell layer just under the bark to be combined with the sugar and carbohydrate blend the leaves were creating from the sun and the carbon dioxide in the air.

❖ Some of the tree's "blood" was then sent back down the tree to be stored in the roots; the rest was distributed between the many knobs hanging from the branches.

❖ Take a close look at your apple.

❖ Is it darker red on one side?

❖ That side was probably facing out and got to bask in the sunlight as it grew.

❖ Is it misshapen? Maybe there was another apple or branch crowding it as it grew.

❖ Are there any imperfections in the skin?

❖ These might suggest a rainy stretch where the apple grew quickly using the resources at hand.

❖ Or it could be a sign of a hail storm, or where the wind caused it to bump up against a branch regularly.

❖ Your apple continued to grow through all of the struggles happening around it.

❖ In October, when the nights became cool, the meat within the newly formed apple changed.

❖ A chemical reaction was triggered by the warm sunny days and the cool moist evenings.

❖ Until now, the apple would have tasted bitter.

❖ This is a defensive mechanism to keep animals (including humans) from eating it before it is completely ready.

❖ Now the apple meat turns sweet, sending out a subtle scent to animals passing by.

❖ The apple you are holding is another gift from the tree to its neighbors in the animal kingdom.

❖ Within the apple is the real treasure.

❖ The tree has produced three to five seeds and surrounded them with this enticing sweet flesh.

❖ Smell your apple.

❖ Can you smell the Earth?

❖ Can you smell the sweetness of the concoction that the leaves produced and fed to the fruit?

❖ That is the smell of sunshine and rain.

❖ Does that smell trigger your stomach to growl?

❖ Our bodies are designed to recognize the nutrition within the fruit and to desire it naturally.

- ❖ When you hold your apple, feel grateful for all of the work the tree put in over the years, and specifically in the past ten months to create it.

- ❖ This apple was grown just for you.

- ❖ The tree wanted you to take this apple and eat it.

- ❖ All of the vitamins, minerals, fiber, and natural sugar have been constructed into a delicious, juicy item created for one purpose only, to supply you nutrition.

- ❖ The energy the tree used to make this apple cannot be returned to the tree except through sunlight, water, and soil.

- ❖ As with any gift born from a place of true generosity, there is a feeling passed between the giver and the receiver.

- ❖ Take a bite of the apple and chew it slowly.

- ❖ As you eat your apple, pay close attention to how you feel.

- ❖ Savor the juices.

- ❖ Let the flavors explode on your taste buds.

- ❖ Taste the effort that formed that apple.

- ❖ Feel how your body's hunger is satisfied in a way that is hard to come by with processed imitation foods.

- ❖ An apple tree created this perfect apple for you and gave you a piece of its life energy that you can feel.

- ❖ But what does the apple tree want from you as a partner?

- ❖ The tree produced hundreds of seeds and packaged them in tasty apples like the one you are eating.

- ❖ The tree is hoping that one or two of its seeds will end up in a little patch of moist soil somewhere where it can grow into a new tree.

❖ Just as the tree offered food to the various insect families for their help in pollination, the tree offers you sustenance in exchange for dropping those seeds somewhere new.

❖ Don't feel guilty if the seeds in your specific apple don't go straight to the ground.

❖ The tree understands that not all of its seeds can grow and accepts that.

❖ Plants understand balance at a far deeper level than we do.

❖ Simply look at a natural wilderness area and observe how each species works together to support all of the life within the system.

❖ No one plant dominates; each brings value to the others.

❖ As humans, we have a responsibility to ensure that at least some of the seeds from our apples get a fighting chance.

❖ We can do that in many ways, including through composting, gardening, working in the community to help build and maintain green spaces, and supporting local farmers who nurture food plants in sustainable ways.

CHAPTER 3

FOOD AND HEALTH

When considering food and food systems (based on the countless conversations I have had while researching this book), people usually feel the most confident about their knowledge and understanding regarding how food impacts their health. Terms such as vitamins and antioxidants, diet programs involving low carbohydrate, high protein, eliminating sugar, counting caloric intake, vegetarian and vegan preferences, and measuring physical activity with wearable gadgets are all common place in our vocabulary and culture. When I ask someone about what is important to them with regards to food and health, the answers almost always trend toward what actions they take toward limiting or focusing on a specific food group as the one thing that will improve health. The ability to respond using the most current lingo initially conveys a sense of depth of understanding and consideration regarding all the facts. Yet when I ask follow-up questions, I often find that familiarity with nutrition and health terms does not actually translate to well-rounded comprehension of a long-term balanced meal plan.

Digging deeper into how we interpret health reveals that there is not one concise motivation for our actions, or how we perceive the actions of others with regard to food. For health-conscious individuals, when I ask how important "whole foods" are in their diet, I regularly observe their eyes darting from side to side as if I have called them on a bluff before they answer with the nutrient group or diet restriction most important to them. Being vegetarian or vegan seems to be perceived by both proponents and opponents as tied exclusively to moral and ethical values, rarely involving a thoughtful explanation of the health benefits of a meat-free diet, and the importance of compensating for the lack of fats and protein necessary for balanced nutrition. I cannot find a single instance in my notes of someone simply stating that they strive for a balanced diet by eating a wide variety of whole foods, even though variety and whole foods are the best way to ensure proper nutrition.[2,15,60] We may think we are eating healthy, but research shows that both the total amount and variety of fruit and vegetables in our diets falls short of what is needed.[60] Everyone, including myself, tends to identify themselves by what they avoid (versus include) in order to foster better health.

It makes sense initially that a mentality of avoidance and limitation are appropriate when observing how our communities struggle with the rising overweight and obesity crisis, but withholding foods based on culturally promoted "dieting" trends and categorizations of "good" and "bad", along with a focus only on the physical body, is dangerous. How food impacts the health of each individual is important beyond this reductionist way of thinking. Emotional and mental health are directly related to diet,[2] and health of the environment and community impact the ability for each individual to maintain their overall health.[16,17] Even from a more selfish viewpoint, the desire to protect a space in which one can thrive should be a basic instinct. Health of the entire food system is impacted by the actions of each individual, and the quality and quantity of food available to every individual is determined by the health of the food system. For these reasons, it is important to consider that while we may feel confident in using food health language fluently, there could be more to learn, and that information may make it easier to achieve our health goals.

Like defining the term "food" as whole foods with minimal processing or additives as discussed at the start of this book, to proceed here the term health also needs to be defined. Health refers to the physical body's ability to move around avoiding injury, pain, and disease as well as the brain's ability to process information and regulate emotions, *and for all those systems to work in concert.* Too often the impact of food on mental and emotional capacity is ignored despite the medical community's solid understanding of this link.[2] Trained, registered dietitians/nutritionists (RDN) educate their patients that no single nutrient, be it vitamin, mineral, or antioxidant, is the key to health. A balance of all the components found in food including carbohydrates, fiber, fat, and protein play key roles in keeping our bodies and minds working in synch.[2,15] The correct proportions and balance between these components, and the amount of physical activity paired with them determines health status. One's "diet" or regular choices for consumption may not be exactly the same as another's based on their preferences but does need to include all of the various substances found in food to function at the highest level. A deficit or excess of one or another component will impact and degrade the entire system even if the effects are not noticed immediately. This gradual weakening, subtle symptoms that can be ignored or excused, are the most dangerous because we tend not to relate them to our overall "health" or treat them with a targeted set of meals to correct.

Because RDNs can be found in most locations in our country,[2,15] and they understand the importance of balanced nutrients acquired through a variety of whole foods for physical, emotional, and cognitive health, one could expect the general population to be savvier. The truth is that the expert RDNs' knowledge feels like a secret waiting to be discovered, and there are several reasons why. A multibillion dollar "fitness" industry, and a separate multibillion-dollar advertising industry both so large that the consumer sees countless messages a day promoting "single ingredient" news, and an overwhelmed health-care industry that can barely keep up with acute health issues are some of the leading distracters.

Common sense might suggest that a growing fitness industry would positively impact and reduce the number of people depending on the health-care system to treat heart disease, diabetes, and other weight-related conditions, and yet in the United States, more than 42 percent of adults and 18 percent of children are obese according to the Centers for Disease Control and Prevention (CDC),[61] and these numbers are growing! We need to reduce our national weight per person and transition to a "balanced diet" to improve our health, but what does that mean? Balance is defined as a state of equilibrium,[62] yet the most common approaches to diet are to restrict or avoid certain foods, eat specific foods in excess for their advertised benefit, or eat what is most readily available regardless of its nutritional value either by choice or necessity. None of these actions are focused on equilibrium or balance, and the result is a population in prolonged crisis mode. The stage is then set for the advertising industry to offer countless remedies based on convenience and ease rather than wholesome balanced nutrition.

Unhealthy eating habits have developed over generations, not overnight; a significant portion of the population has grown up with food being many things: entertainment, competition, a commodity, a chore.[9] Rarely is food considered an opportunity to enjoy a healthy and productive life experience; it is what must be consumed to survive so we can move on to earning enough money or finding enough time to have that life experience. Surveys show that food choice is primarily driven by taste then by price; how food contributes to healthfulness is an afterthought.[63] Our priorities don't value food, and as a consequence, the health of our whole being is eroded. While "foodies" may be passionate about certain special meal experiences, how many people clear their mind to revel in a peanut butter and jelly sandwich or give praise for a quick apple between tasks? Our current interactions multiple times a day with food are not noteworthy or precious, they are necessary and routine. This relationship with food has made us susceptible to and desperate for miracle cures because we do not recognize that eating a balanced diet is the best medicine.[64]

Faced with marketing in literally every facet of our daily lives from advertisements on electronic devices, web sites, and social media, to product placement in entertainment, to diet groups in the workplace, the options for quick fixes are endless and easy to access. Advertising of any kind is based on the principal of making something desired. For a population that already has a desire to feel better and improve their health, being offered products that claim to meet that need without much effort required is extremely attractive. On the other hand, making appointments to meet with a RDN regularly and/or doing the hard work of changing shopping, meal preparing, and eating habits is not. But like anything, the value you get out of something is equal to the effort to put in. Perhaps partnering with a RDN to plan improvements is not necessary, and change can be inspired independently, but having a resource for encouragement can be a valued asset and provide accountability. The great news is that many health-care plans offer access to these professionals who specialize in this preventative medicine for a living! Check with your health-care plan provider to determine if this option available is to you.

Both diet and physical activity are important parts of a balanced health plan,[2,65,66] and the marketing of exercise is big business. The impulse to disregard changes to what we eat and instead try to "fix" our health with workouts that can be tailored to specific time windows in our existing schedule is strong. The for-profit fitness industry including gyms and personal trainers, home exercise equipment and programs, reality TV weight loss shows, and infomercials, utilize the advertising industry heavily to project a concept within our culture of what fitness and body image should and can be. The reality is that achieving even close to what is considered the "goal" takes a great deal of time to attain and maintain as anyone with a gym membership can attest. Physical health can be improved through structured exercise using fitness equipment, but once again, the topic of balance becomes a question and could be part of the reason why "working out" is so hard to develop into an enjoyable and sustainable habit. Structured fitness focuses on the physical body but tends to ignore the brain as an integral part of wellness. The mantra of the fitness industry is often: the body is the problem, food contributes to the problem, food can and should be replaced with nutrition substitutes designed to boost the exercise routine success.

If working out worked that way, then it wouldn't be so hard to maintain as a long-term habit. Instead, exercise and food must work together in a balanced way to deliver premium results, and enjoyment of both activities is optimum for physical, emotional, and mental health.[65,66] What does a balanced perspective look like? Consider for a moment what language you use every day in regard to food and exercise. Are the words "I can't," "I shouldn't," "I have to" an everyday occurrence? Are positive phrases such as "I enjoyed," "I'm excited to," "I can't wait for" only uttered on special occasions? The attitude and language you approach food and health with could be an influencer of enjoyment but could just as easily be a result of an imbalance. In fact, multiple studies have shown that less structured, strenuous, and repetitive physical activities involving nature help to improve physical, mental, and emotional health equally and are easier to establish as an ongoing habit.[16,17] They also help increase appetite, a sign of the link between fitness and food. Taking a daily walk in a park or garden doesn't cost much though, so there is less of a push by advertisers to encourage that type of activity over the purchase of a device or membership which is likely to be discarded and replaced by a newer product each season that promises to produce better results.

To compliment the fitness industry and provide an alternative for those who don't have the time or inclination to push themselves physically, other industries have blossomed over the last few decades. The for-profit "nutrition" and dietary supplement industries have been developed by advertisers to promote the validity of products backed by "science." This is another example of how food must be considered with an interdisciplinary lens. The relationship of food and health is intimately intertwined with science, politics, and the substantial economy of the industrial agriculture food system. Contemplating all the different food perspectives and their complex interrelations is frustrating; I have literally spent years actively researching the subject, but to ignore the intricacy is to deny that our lives depend on food in multiple ways. Beyond that, I believe the quality of our life experience depends on how we perceive food on several levels.

Allowing for whole foods, imitation and substitute foods, and synthetic food replacements to be marketed indiscriminately or with elevated status for manmade substitutes forces us to forget the sophisticated and sometimes mysterious ways that whole foods provide perfect nutrition.[2,60] In humanity's urgency to discover through science, we sometimes destroy without consideration of what is yet to be learned. *Let me state clearly that I firmly believe science is vital. It brings value to humanity; it can provide us the tools to understand our world and how to protect it for the benefit of all living organisms.* The problem is that we humans have a habit of reducing what we don't know into bite size pieces to study it, and forget to apply what we learn to the larger whole once we feel we have exhausted the unknowable.[48] With regards to seeking healthy food options the consumer must be cautious of products backed by "scientific" study as the information may be taken out of context, or not have passed the rigorous peer reviewed vetting necessary to be considered validated. Philosopher and author Ruse explains quite eloquently how the scientific community, although guided strictly by process and testing can have their work misinterpreted and manipulated by power, money, and prejudice.[48] Pure science is driven by curiosity and confirmed with evidence; pseudo-science uses the technical language we expect to promote outcomes that do not meet the high standard of peer reviewed experiments.

Today in the United States, we are seeing the impacts of what happens when a significant portion of the population begins to doubt science because of the prevalence of pseudo-science infiltrating various media outlets without any system to identify or control them. The tug of war between real science and misinformation dressed up to look like science is literally costing people their lives as we battle a global pandemic. For the consumer, the old saying "if it sounds too good to be true, it probably is" can help trigger red flags. Claims that a pill, powder, or any one nutrient can provide fantastic results should be rejected and ignored, yet consumers spend millions of dollars on quick fix trends every year rather than simply buying and preparing whole foods that provide balanced nutrition.

One of my favorite questions to ask friends, family, even casual acquaintances, is "what is your biggest hurdle when it comes to eating better?" Most of the time the answer is that good food is too expensive, yet about half then explain either fitness or nutrition aids they purchase to make up for the less nutritious food they eat. As an example, recently, I was speaking to someone during a lunch break who noted that organic chicken was so much more expensive than conventional industrially raised chicken when they were grocery shopping last night, and they couldn't afford it in their budget. I then asked what they were drinking as their lunch meal and they told me all about the special protein shake they have for lunch every day as part of their new diet. I am curious if the irony is as obvious to others as it is to me. It seems like purchasing the high-quality organic chicken and portioning appropriately would reduce calories and supply clean protein, and ultimately be less expensive than buying cheap meat, the protein shake mix, and the items used to dress up and hide the flavor of the shake. Definitely ironic.

Ultimately science is not to blame for the many pseudo-scientific claims promoted daily to sell products, but the scientific community has allowed an opening for health and nutrition advice to be more easily co-opted, manipulated, and mistrusted. Nutrition discoveries, the investigation of the various components and reactions that occur in the very complex workings of the human body, have been and continue to be studied on a single nutrient level. When the importance of that nutrient and its benefit is confirmed, the announcement often highlights the new information without providing appropriate context and reinforcing that dietary guidelines should promote variety and plant based whole foods as the most important factor for health.[2,15,60]

Here is an example of how science is manipulated for profit. The process goes something like this: Scientific research discovers that antioxidants help prevent cancer! The for-profit nutrition industry bursts with countless product options for the consumer to get their man-made antioxidant supplements. Follow-up information shows that antioxidants are most effective when consumed as part of the complex compounds found in the whole foods that contain them (like blueberries, for example). The nutrition industry might begin to offer a few blueberry-flavored versions of their antioxidant products, but for the most part, the consumer has moved on to another discovery feeling confident that the synthesized antioxidant product has improved their health by reducing their risk for cancer. The number of cancer cases diagnosed annually continues to grow.

If scientists took more care in framing their discoveries in context to the larger nutrition picture, if media held to a promise of contextualizing instead of sensationalizing, and if advertisers weren't so eager to push a product for profit, perhaps consumers might not be as quick to pounce on the new latest and greatest breakthrough with little overall improvement over time as evidenced by the declining health of the nation. The term "antioxidant" itself is proof of its commercialization. Polyphenols are the chemical compounds that produce the antioxidant effect, but "polyphenols" isn't as fun to say so the consumer has been taught the easier word to remember. Ultimately, polyphenols (antioxidants) are a crucially important factor in a nutritious diet, but cancer has not been eradicated because of that knowledge. The message that needs to grab the consumer's attention is that a balanced diet of whole foods eaten over a lifetime provides the body the tools, including polyphenols (antioxidants), to fight the contaminants that we encounter as part of the world we live in.[2,15,60,64,65] Drawing attention to contaminants in any sense, but especially in relation to the food system is especially taboo in marketing though. Yet this is another important part of understanding how to choose the foods to best support physical, mental, and emotional health.

Contaminants in food take on many forms. There are multiple dangerous bacteria that can be introduced during the growing, harvesting, processing, and preparing of foods.[2] These are monitored and regulated by some global and federal government policies to help protect the health of consumers;[8] note the interdisciplinary connection. Other chemical contaminants that are intentionally or unintentionally introduced to the food system are another topic altogether. Additives, preservatives, and food packaging are usually assumed safe as the government has processes to keep tabs on corporations.[8] Large-scale acute illnesses are uncommon, so the system works as designed for the most part although it is not perfect. Contaminants in food packaging (and other chemicals used to preserve foods) can cause a slow buildup of toxins in our bodies over time.[67] Long-term impacts of how these contaminates impact our bodies are rarely considered though, because we are exposed to such a wide variety of toxins in micro doses that rarely trigger acute sickness.

Perhaps the most debated controversy surrounding chemicals in the food system stems from those used in conventional industrial agriculture where negative impacts on the health of individuals, the community, and the environment can be observed more clearly. The development and use of pesticides and herbicides that proponents and the government have claimed for over a half a century are necessary to produce enough food to feed the growing global population might seem like old news, and rightly so as questions about safety have been around for many people's entire lifetime. Carson, a well-known scientist of biology and ecosystems, painstakingly documented and exposed serious health impacts on both humans and the environment for chemicals used prolifically in the twentieth century,[68] sometimes for no reason other than they were available, and the value of multiplicity of species in the ecosystem was not generally understood. Her work was monumental for raising awareness in the public perception, yet today, nearly sixty years later, pesticides are still being used,[69] still being found in our food,[15,36,70,71] and their accumulation in our bodies are being explored as the root cause of the distressing increase in nervous system diseases today.[72,73,74]

Because some scientists produce studies that the government uses to regulate pesticide use as safe for the community in certain applications, while other scientists are demonstrating the hazards of long-term exposure to those same chemicals, it is understandable that the lay person who comes home from work tired and has to make dinner, help children with homework, and handle daily chores around the house isn't then going to want to read scientific articles before they go to bed so they can make an informed decision who to believe. How does this relate to each individual's interpretation of information regarding nutrition? Consumers become susceptible to fear as another motivator for buying the latest product supported by a pseudo-science study. Out of necessity, they are also forced to trust that political leaders and government agencies have their best interest at heart, and that steps will be taken promptly to address concerns as they develop.[8]

This is another example of interdisciplinary cross over that is unavoidable with regards to our food systems. The problem then falls to the government to raise funding for unbiased, thorough, detail-oriented studies about the safety of food, and the chemicals now found in those foods. As that funding is extremely hard to come by because no one wants their taxes raised, the studies produced and paid for by the various industries making a profit from the sales of chemicals and highly processed food replacements become the resources used to guide regulations.[8] The private sector investment to fund quick and sometimes sloppy pseudo-science, which floods media sources, pays off at the consumer's expense; negative impacts from these products may not be understood or confirmed until overwhelming evidence forces new studies to be commissioned at some point later down the road.

Contradicting messages about our food system are not limited to just nutrient promotion over a balanced diet, the safety of food despite man made contaminants, or the necessity of chemicals in food production. Nationally obesity rates are rising, as are hunger and malnutrition, and these two seemly opposite conditions are impacting the same groups of people at the same time.[2,6,75] Proponents of agribusiness technology have no problem explaining how crucial it is to produce more grains to feed a growing global population and then turn around and boast about the amount of grains currently grown per acre that provide well over the daily requirement of calories per person needed to be healthy.[41,42,75] Yet another example of so-called experts promoting a lack of balance as a healthy diet should include whole grains, but not be dominated by them.[2,15,64,65] Supporters of aggressive agribusiness technology growth are dangerous in another way, too; they chip away at the health of the community and overall food system though systematic messaging that becomes accepted as truth and influences food choices. Defending the need to increase agricultural production with chemicals is often backed up with a dismissal of alternative options. A perfect example is organic growing methods, production rates, and quality of final product.

The statement that organic growing methods can't produce at the same rate as conventional farming is an accepted principal by many despite the fact that it is entirely false.[21,32,33,36] What is true is that farming organically requires a different process than simply using chemicals, and farms new to organic growing may need a couple of seasons to improve the soil adequately and reap its potential. While this may seem to be unrelated to our personal health, it is intimately linked through the perception that healthy food is limited and therefore only accessible to those with financial means. This accepted mindset impacts poorer communities and stifles desire for nutritious whole foods which were barely affordable or accessible for these communities to start with, confirming the perception of exclusivity.[6] Without demand, supplies disappear, and food deserts are formed. This stimulates consumption of low cost highly processed imitation foods full of empty calories and an increase of the addictive physical reaction experienced when too much processed sugar and unhealthy fats are consumed on a regular basis. Lack of access to healthy food both contributes to and prevents communities from escaping poverty. Once again, the link between mental and emotional health emerges as equally important to physical health, as a poor diet contributes to depression and other mental illnesses that plague poorer communities.[6,75]

The debate over organic versus conventional farming is often framed as a topic for the educated and financially comfortable in the population to banter about without much impact on the portion of the population who needs access to healthy foods more than ever.[46] For poor communities, any supply of fresh vegetables (organic or non-organic) would be better than the current situation, but lack of demand due to cultural norms promoting imitation food doesn't provide suppliers any encouragement to make the effort to deliver consistently.

Adding to the debate of organic versus non-organic, GMO (Genetically Modified Organism) foods trigger a much stronger reaction with the population and have dramatic consequences either way. Proponents become even more insistent that there is absolutely no way to feed the growing population without GMO foods, pointing to populations of people suffering from malnutrition and starvation in developing and poor countries as the primary benefactors.[41,42] But these are not the markets growers plan to sell their products in, and those hungry nations are not being fed with the surpluses of grains already grown;[30] exactly how would growing GMO foods change that political dynamic? Ultimately, there are two truths about GMOs with many other supporting factors stemming from each. The first truth: GMO foods use technology to reduce the need for some (but not all) chemicals and increase production of single "monoculture" crops.[76,77] The second truth: Many people consider GMO technology to violate certain ethical principles and believe it is too soon to understand with certainty if the technology is safe for humans or the environment.[21,28,36,37,39,69]

Proponents of GMO technology state that the act of merging genetic material from completely unrelated species in a laboratory is no different that the centuries old practice of intentionally cross breeding plants and animals to draw out desired characteristics. This mentality is deeply rooted in belief that humans are superior to all other life forms. The very real difference between cross breeding and genetically modifying a living being is that cross breeding allows nature to reject the potential combination. Cross breeding takes time; multiple generations of the species being adapted before the changes become inherently reliable. Genetic modification allows the manipulation of a species well beyond what nature would ever allow with no generational testing or observation. The product is then pronounced safe to ingest simply because it was created.

No amount of protest or demonstration can convince opponents of GMOs that the benefits outweigh the risk, and history is full of past technological breakthroughs originally considered safe, that time has eventually shown to be dangerous. (Consider the studies cited earlier in this chapter that note possible links between pesticide build up in our bodies and nervous system diseases like Parkinson's, and the possibility for that build up to be passed to future generations through gestation after nearly a century of use that is still considered "safe.") The motives of those fighting for GMOs are also distrusted as their arguments for said technology seem to be a thin cover for the profits that a small few stand to gain if successful. The ethical argument is considered an unnecessary and frivolous roadblock by supporters, but isn't that by itself cause for concern? Science must be bound by ethics, especially in relation to the food that feeds us. Ethics are not a luxury to be exercised when convenient; they are the guiderail that prevents humanity from destroying itself and everything around it.

I have read more articles and opinions that promote GMOs than I ever would have if not for writing this book. I wanted to be as completely informed and objective as possible rather than make a snap judgment. What I learned is that pro-GMO information can always be traced back to the source of those who stand to gain from it in some way or those who consider technology to be more important than everything else and at the expense of everything else. To me it is like futuristic movies that warn that Artificial Intelligence will eventually turn on humans and destroy them because humans failed in some way to account for humanity as valuable when creating the logic to guide A.I. When reading about GMO benefits, I always felt a little sick, and I trust that gut reaction even if I don't understand exactly what is causing it. I know enough about the subject of food to state that I believe there are other options for feeding the global population. Those options may force our food system to develop down a different path than we are currently headed, but in the end, it has the potential to be healthier than the system we have now; an improvement. If science and technology were driven by ethics rather than profit, and feeding those who are desperately in need was a mission and a passion completely divorced from profit or power, I do not think there would be an increase in hunger. Quite the opposite; I would be willing to wager that hunger all over the world would be satiated, and what it means to be healthy in a balanced way would be inherently understood more deeply that is imaginable today in the United States.

The GMO discussion set aside, we do have some facts about the nutritional quality of organics which can help us decide what to eat today and tomorrow while we ponder the ethical quagmire of global health and nutrition that can't possibly be resolved entirely this week or this month. To date, the number of studies focused on the nutritional differences between organic and conventionally grown produce is limited, and some of those studies lack certain protocols that are considered the gold standard in scientific testing. Still overall, the studies that have been done all point to the general outcome that organics have significantly fewer chemical residues in their tissue.[70,71] As nutritional biologist Nestle puts it, if the consumer is worried that chemical residues are a contributing cause of illness, they should strive to choose organic options whenever possible.[15]

A less clear result of these studies, and something I really wish would be studied more, is if organic options actually contain higher overall nutrient content than their conventional counterparts. The studies available do show that there are differences in nutrient contents, often higher in organic samples, but not always. In part, this could be due to the specific nutrients focused on in each study, but without a larger more comprehensive investigation, it is hard to know. As a side note, these studies confirm, and many sources agree[21,37] that nutrient content begins to diminish from the moment produce is picked so the fresher the specimen, the healthier it is. While a thorough understanding regarding the differences in nutrient content as a result of growing process may still be in the investigation phase, perhaps basic observation can hint to what will be found.

Here is what I know from my gardening experience over the past twenty-five years. Vegetables may grow and produce even if the soil they are in is deficient. When a soil is rich in organic matter (compost), the plants growing there will have deep rich color in their leaves and fruit, they will produce unbelievable amounts of fruit, and their fruit will smell and taste unimaginably better than any store-bought item. From basic observation that tells me that the plant grown in the healthy soil was able to absorb all the nutrients it needed to produce its fruit, and that fruit as a result contains those nutrients. I know from experience that a plant in deficient soil that is given a fertilizer high in nitrogen will likely have rich color in the leaves but may show other evidence of deficiencies like reduced fruit production and reduced biomass (the overall quantity of stems and leaves). Improving soil quality is the best way to ensure healthy plants; it is more effective than fertilizing with chemicals. This evidence convinces me that organic vegetables do have higher and more naturally balanced nutrient content even if I do not have a lab to prove that in.

Armed with the knowledge that the best nutrition is to be found in fresh whole foods grown as cleanly (avoiding chemicals) and eaten as freshly as possible, and that balanced nutrition nurtures physical, mental, and emotional health for a whole-body wellness, we are ready to consider health beyond our self-centric perspective. To grow the foods that are most nutritious, we need to be concerned about the health of the environment.[13,21,28,30,36,37,44,49,69,78] A landscape that has been stripped of animals, insects, and plant life will not have healthy soil because the mechanisms to recycle nutrients through the system have been removed. A barren land has no nutrients to share. If the environment is sick, lack of nutrients is not the only issue for human health either. Our mental and emotional health is tied to not only what we ingest orally but also how our other senses are fed.[13,16,17,21,28,39,44,69,78] The air we breathe is recycled by vegetation. Our eyes and ears are soothed and stimulated by natural scenery. We need interaction with nature on a much deeper level than most of us currently get if we are to experience comprehensive health. We may be able to "survive" for a lifetime on a bare minimum of basic nutrients, and many people likely do without ever knowing what whole-body health is. Wellness is like awareness though, once gained it is hard to ignore.

The fact that each person is on their own health and wellness path contributes to the different perceptions of value applied to food and its importance to the individual, the family group, the community, and the nation as discussed in the last chapter. Those with a higher level of food health understanding are going to not only want to share information with their family, friends, and neighbors out of the simple desire to see their lives improve but also ensure that everyone's access to food is protected. This outreach may be rejected for various reasons including, inability to visualize the bigger picture, confidence in conventional accepted norms, or inability to comprehend based on an unknown health imbalance. Issues like how a largely unhealthy population costs every citizen when it comes to health care, and not only in insurance premiums but also in ability to maintain some form of insurance, lack of substantial preventative care options, abuse by the pharmaceutical industry, and gloomy resignation that there is no way to change the system as a whole, can leave us feeling powerless.[6,8,46] As a result, we may recoil from advocating food health proactively or receiving solution suggestions from others even though it is for the benefit of all.

The interdisciplinary relationship of food and health with politics, economics, and science further complicate and discourage the urge to promote it as the topic becomes one that foments adversaries rather than compatriots. I have experienced this myself on many occasions with friends who become frustrated by the complexity or defensive about their accepted understanding when it is challenged by a new perspective. Thinking about food health from a top-down viewpoint is self-defeating and triggers resistance to anything that may disrupt the status quo; fear of change is an emotional hurdle that must be addressed before the debate of food can even be had. Moving forward, it is possible to argue that our culture could continue to use conventional and technological agriculture to support the population, maintaining the level of degraded health we accept as the norm, until some cataclysmic event such as war, or series of climate induced super-storms, or other geologic event forces a significant population adjustment. The alternative is that individuals stop and think about the food that supports their health differently, so differently that change is desired and sought after. Evidence from communities with strong food systems demonstrates that this trickle-up model produces real results, often quite quickly.[79]

How should we change the way we look at food? What this looks like for each person is unique although there are some general themes that apply to most. Before taking any action, every individual should evaluate honestly what their current attitude toward food is. Is food bothersome, unfulfilling, something inescapable and un-enjoyable, a crutch? Is food (or the lack of food) something that consumes everyday thoughts? Is food a lifeless "thing"? If any of these phrases summarize food accurately for you, consider the types of food (whole and fresh, imitation, highly processed, synthetic) that make up your diet. Can you even identify which categories your normal food choices fit into? What proportions of your meals are "alive"? Perhaps considering whole foods is not new, but the idea of food connected with life and energy is. Does thinking about food as living, filled with energy, and part of an active exchange of energy, feel completely foreign or creepy?

As a gardener, I enjoy spending time around the living plants in my "edible landscape." My flowers and food plants all grow together in what is called a polyculture. Watching how various plants work together to support, protect, and aid each other, as well as provide for me helped me to define food differently in my life; it was this experience that promoted me to start making changes in my own diet. It stimulated desire for foods that look natural and taste fresh. Instead of forcing myself to eat something "good" because some media authority has told me all about its benefits, I nurtured my own curiosity about the intent of the living being that grew the whole food item for me.[80] Some might think this is a stretch, but most plants bear fruit designed to attract animals that will consume it and spread its seeds. The most basic and natural interaction between the plant and animal world is one of exchange for the benefit of both parties.[64]

Eating meat doesn't mean abandoning this appreciative exchange of life force mindset. Many Indigenous people all over the world provide a model of how respect for an animal food source can be preserved and accepted as a gift.[13,28,29,30] In the United States, some deer hunters enact a modern version of this partnership, respecting and consuming the meat provided by the living being who has surrendered to them. Small farmers who raise "traceable" meat from livestock that they care for and nurture, who respectfully harvest those animals with gratitude are another example of how to approach food as living and precious. All life is part of a cycle; at every level, one being gives to another in an exchange. The system only becomes unbalanced when one side takes without awareness, consideration, or restraint.

Shifting perspective, realizing the difference between what we have been eating without thought and whole foods bursting with nutrition meant to keep us well can be truly uplifting. Desire for whole foods is cultivated delicately like starting a campfire with leaves and twigs and blowing on it gently. As the fire of desire grows, making new choices for the next meal or snack doesn't seem like drudgery or an obligation. Instead, we can look at each opportunity to eat as a moment to be pampered. Looking at food (whole fresh clean food) as an opportunity to break free from the monotony, to revel in a full body celebration of the taste of what it means to be healthy, encourages us to continue to make changes that provide nutritional balance. Eating regularly like this reinforces the smart choices with positive physical benefits. A mindset of awe, gratefulness, and appreciation also begins to impact our emotions and general outlook. It becomes harder to see the world as a cutthroat competition bent on eliminating us. Of course, there are still challenges; a co-worker's annoying habits or a customer who is too distracted to be polite. But the opportunities throughout the day to be rewarded and protected with whole food snacks or meals are plenty. Enough to tip the scales in the favor of positive reward and partnership with a team of foods that literally fill you with the light energy they have collected from the sun.

At this point, you may be thinking something like "I was with you right until that last part" but consider this: professionals have known for a very long time that a balanced diet of low-calorie nutrient-dense vegetables, healthy fats, protein, fiber, and complex sugars are what we need to be healthy. In fact, we all probably know that at some level. But our nation is sick, and we spend BILLIONS of dollars a year to fix our ailments between nutrition supplements, diet plans, and fitness programs;[8] we chase after health never catching it. Continuing to approach food with our current attitude as if it prevents us from enjoying a healthy life is not working.

What is the harm in choosing to think about food differently? If it inspires us to change our diets for the better, does it matter if it seemed silly before we attempted it? It may feel awkward at first, but no one has to know what we are thinking; thoughts are private. It could take some time to discipline ourselves to a new food attitude, after all we have been thinking about food as the enemy for far too long. A helpful way to get started is to grow something, anything, edible. Herbs or salad greens in a window pot, green beans along a walkway, strawberries in a hanging garden bag, a tomato, pepper, or eggplant in a garden. Providing for something that then provides for you is a great way to develop an appreciation and adjust mindset.

The next step after thinking about food differently is eating foods that fit the definition of being alive. Whether you visit a farmer's market, check out a CSA group, or simply spend more money in the produce section of your local grocery store, it is important to search out the freshest alive or whole foods. The primary reason is to discover what freshness tastes like and consume the highest nutrient levels possible. How our bodies react to foods is subtle, especially if we are slowly transitioning to whole foods; picking fresh high nutrient foods will give us the best change of recognizing how those foods make us feel while we are eating them, digesting them, and using the energy they have gifted to us. Choosing fresh means choosing seasonal; if you are not sure what is seasonal in your area, a farmer's market is a great place to learn. At a grocery store, look for tags that say local; this is a selling point so it will be noted. In the northern regions, January and February can be tough months to find local foods so look for regional or nationally grown foods. When seasonal fresh vegetables aren't available, frozen vegetables can be a decent substitute, but try to focus on what you can get fresh and in season.

When it comes to choosing certain vegetables, I simply don't buy them anymore if I can't get them when they are in season in our hemisphere. I love asparagus fresh from my garden. I can taste the strong spring sunlight and rain that force the sprouts to greet me when few other plants have even woken from their winter naps. When my own supply runs out, I will buy from local (the Northeast) or regional sources (which includes from Canada for me). But come June or July when the labels show locations that are further and further away, I will stop buying asparagus. At that point, what is in the store tastes so stringy and bland compared to that in my own garden that I'm left feeling disappointed at mealtime; I would rather take a break from eating it so that the next spring my taste buds are primed for the phenomenal. Popular food writer Barbara Kingsolver's book

Animal, Vegetable, Miracle has some great advice about eating seasonally.[21] Even in the winter months, there are many great options like squash and root vegetables that keep for a long time past harvest.

Finally, make every effort to take note of how you feel when you eat; perhaps keep a temporary journal. Every person is different, and few of us have lived the majority of our lives eating only clean fresh whole foods so our bodies might need a little help reaching optimum performance. Seek out health professionals, such as RDNs, who believe in the power of food to support overall wellness. Advocate for yourself as the perfect balance for you is unique. My experience demonstrates the importance of noting nuance. I eat a healthy balanced diet; I could cut back on total calories a little or increase activity to reach that next level but looking at my diet there are no glaring alarms to suggest deficiency. Despite this, I had illusive but persistent symptoms that something wasn't quite right. Luckily, I had a medical doctor who didn't rush to prescribe pharmaceuticals; she tested my blood and figured out what I needed to supplement with. I noticed a difference immediately.

I point this out because there is a real difference between a supplement recommended by a medical professional based on firm test results, and a supplement advertised as appropriate for everyone and backed by pseudo-science. Alternatively, I have found that opinions run strong about supplements with some RDNs suggesting they are unnecessary in all but the rarest cases. From my experience and research, I think each individual needs to work with trusted health advisors to make that call, AFTER they have shifted their diet and given their bodies time to process the new nutrient volumes.

My personal example doesn't just end with "fixing" what was missing in that moment either. Later when I ran out of my supply, I was lazy and didn't restock, going about my business thinking the problem was solved. Nearly a year later, I realized that those same illusive but persistent symptoms had returned, although I couldn't pinpoint when. I ordered up a new supply, and within a week, I was feeling tip top again. The point is that every person is different, and in this imperfect world, once we have taken the steps to actually change our diet to be full of fresh clean whole foods, we still need to be attentive to achieve wellness for our whole mind, body, and emotional being.

The possible benefits of this mission, in addition to better heath, include a change in appreciation for the environment that food grows in. It can help us to establish values that guide a plan of working WITH the land, supporting, and being supported by a local community.[79] It can eliminate greed and the fear of never having enough so we can see what it takes to make an amount of money to live a comfortable life and view wealth as a measure of joy, not just currency. As more people shift their food health outlook, healthier foods will be grown and available to meet the rising demand. Others who may not be able to access fresh whole foods currently may begin to see their food desert homes sprout with options as communities lift each other up. Whole being wellness, health, is attainable through food.

CHAPTER 4

FOOD AND SCIENCE

I have always had an affection for science; physics was my favorite branch of science in high school, although chemistry and geology were very close seconds. Biology intrigued me; it took a while for me to appreciate it on the same level as the other sciences, but all science was preferable to memorizing historical dates or trying to understand the various reasons for war and conflict in institutionalized history classes. When I returned to school at this mid-point in my life, I knew I wanted science to be the central focus of my learning. I wanted to examine food from a scientific perspective because science has rules and tested processes.[81] What can be learned using science *is proven*, or so I believed.

That is not to say that I no longer trust science, in fact I now trust it more. I have complete confidence that science can help provide understanding about the world we live in, but I have gained a deeper appreciation for what science is and how it should be understood by the community at large. The COVID-19 pandemic has exposed that an uncomfortably large segment of our population harbors a mistrust for science. I am truly saddened by this fact and admit I don't fully understand how science has become more of a belief system than the process by which to explore our curiosity about how the world around us works. Perhaps it is because profit-driven businesses have imitated the processes and technical language of scientific study to push products on us that erode our confidence.

I have struggled with how to delineate between the pure and exciting study of science I admire and the tainted concept of science that most people only experience through media and advertising. Dr. Linda Jones, a Professor at Empire State College whose passion for science is inspiring, helped me understand that authentic scientific research and discovery is alive and well. Recognizing it simply involves taking note of the integrity and candor of the objectives and theorized discoveries that are the anticipated outcomes of a project. Why is something studied? What is the expected result of the study? If an unexpected result occurs, how do the scientists handle it? I now pay close attention to the language used regarding science and remind myself regularly to question whether the evidence available matches the conclusion.

Science is best described as is a "culmination of what we "know"/ have learned about the biological, chemical, and physical world through observation."[82] The tools used by scientists during research provide a way to systematically and consistently test ideas and present them to others for verification. "Science is a snapshot"[82] of how our world fits together and functions that is dependent on what is known at the time; it is "derived through observation and driven by perspective."[82] Science is a way to help humans explore curiosity and answer questions, yet today, it often results in an accepted (assumed final) knowledgebase that tends to smother our naturally inquisitive nature and resists the challenge of questions that don't fit into the accepted *laws* that have been created.[48] Biologist Haskell writes, "(S)cience, done well, deepens our intimacy with the world. But there is a danger in an exclusively scientific way of thinking."[83]

Science can only validate, determine, argue, illustrate, or support a theory in relation to the collection of other ideas that inform one's world view. Science is an "observation-based system."[82] If you accept what has been hypothesized, tested, and agreed upon by experts in a given field, you will approach new scenarios with a perspective informed by that established idea. Science is at its best when the scientist asks hard questions that are sometimes unpopular or contradict what is "known" and then tests those ideas rigorously and actively participates in the peer review process. Scientists should identify unknowns and explore them with as little bias possible from every angle, until both the details and the larger perspective provide us a greater understanding of our place and purpose.

Understanding science in this way helps explain why knowledge is constantly evolving, and accepted concepts "shift"; they are set aside after a long period of functioning as the foundation for various other conclusions.[81] Scientific discoveries or breakthroughs are common and necessary as the collective human intellect grows. Importantly, as specialists or experts in a discipline explore details within their field, it is crucial to regularly check in with other disciplines to assess whether an idea withstands challenges in the context of the larger picture. Looking at a puzzle from a different perspective can shine a light on whether an assumption appropriately fits related fields, increasing its validity. There are many correlations between practicing good science and development of an individual's personal ideas and understandings about how all aspects of life should be lived.

Each of us has a world view that is informed and evolves based on interactions and discussions with others. Someone who chooses to interact only with like-minded individuals will develop tunnel vision and their world view will become increasingly radical and unable to be integrated smoothly into the accepted societal understanding, sometimes resulting in a self-imposed isolation and a permanent defensive posture. When this personal stance influences scientific study, the resulting data and hypotheses can be reduced, manipulated, segregated, and weaponized.[48] In scientific study remaining connected and challenging ones' ideas through interactive communication with other disciplines is the measure of integrity; ideas informed through this method can be more reliably trusted.

Applying this proactive approach to the science of food and food systems is especially important because every aspect of the food system is directly influenced by other disciplines; no section of the food system can be isolated from another.

When researching this Food and Science chapter, I expected a strong interdisciplinary connection to the Food and Health and the Food, Politics and Economics chapters, but I was surprised by the significant interdisciplinary relationship with the Food, Religion and Spirituality, and the Food, Culture and Ethics and the Food and Values chapters. Our culture in the United States relies heavily on scientific evidence to support our actions.[48,81] This reliance has allowed for scientific discoveries and studies to be taken out of context, manipulated, and abused, and we as a culture tolerate it because it allows us to find an excuse or *evidence* to defend our actions with. Perhaps most people don't even realize that they are a part of a cultural movement that is degrading the relevance of science by willingly embracing anything presented as evidence without thinking critically about who sponsored it and what the study set out to answer. Emphasizing scientific literacy is a necessity for our society to evolve and improve. I hope this chapter will help refocus all of us on the best way to use science to increase our knowledge holistically and remind us of our responsibility as individuals to continue to ask questions.

When we introduce food to scientific study, two primary focuses take shape: nutrition and agriculture. Scientific study of nutrition has demonstrated there are countless chemical reactions involved in the absorption of energy from what we ingest.[2] These chemical reactions are complex and support one another in ways we have yet to fathom the limits of, but we do understand the basics very well even if the how and why are still being discovered. Nutrition scientists and registered dietitians/nutritionists (RDN) know through observation and testing that a diet of fresh whole foods, consisting primarily of plants provides the human body the best fuel to operate. While our bodies are capable of being sustained on a wide variety of processed items currently considered "food," the functioning of our bodies suffers in subtle ways when we don't eat primarily whole foods.[2] We are biochemical entities. The ability of the human body to filter out toxins that can trigger cancer cells to develop,[74] the capacity of the human brain to process information and regulate emotions and hormones,[16,17] and the immune system's efficiency and effectiveness[2] are all processes that impact the quality of life for an individual, and are directly influenced by nutrition. The extraordinary way in which the living beings we share the planet with provide us nutrition should highlight the importance of understanding and protecting our food sources.[37,50]

Human life on our planet would not exist without plants providing us food. Plants are completely amazing! They create their own nutrition from sunlight, soil, and water and share that nutrition with a wide range of animal species from microscopic to massive.[43,84] While meat from other animal species provides unique nutritional components to our diet, those animals developed their flesh from ingesting additional plant sources, possibly ones we couldn't digest.[2,21,28,37] Consider for a moment how advanced plants are as living beings. Plants have the ability to regulate their genetics to adapt to their surroundings.[69] They have evolved techniques to prepare their seeds for extended dormancy in certain cases; some activate their different sex organs to only produce or accept pollination during short periods ensuring genetic variety from other plants for strength.[30,43] Plants also work together between species to protect and assist one another.[85] An ecosystem has layer upon layer of plant species executing a wide spectrum of processes,[68,86] including producing what we eat for food. Redundancies are not an accident or a result of competition; they are all parts of a larger functioning organism.[87,88] Studies have shown that removing one functioning member of an ecosystem, even when there is another species providing the same service can cause the entire system to degrade slowly, then collapse.[30,68,85]

Thinking about plants as living beings intentionally providing food for their community may seem like radical idea; in the same way, one might criticize my anthropomorphizing of the chipmunk that visits my back deck and spends a great deal of time trying to catch my attention and explain something very important to me with his chirps and barks that I have yet to grasp. Several features of plants validate this perspective in my opinion though. Plants are known to manufacture toxins on demand to dissuade animals from eating too much of them,[84] while others are simply poisonous to most grazers from the start. Conversely, many of our food source plants actually expend a great deal of energy on producing delicious and nutritious "fruits" that encourage animal consumption as a means of seed disbursal.[43,45] Whether prohibiting or encouraging nibbling from neighbors, the ability to adapt their physical processes in relation to the surrounding community and environment has been confirmed repeatedly by science.

In the past decade, scientific study has begun to explore internal actions in various plants with relation to communication, taking the idea of capacity for intent a step further. On almost opposite sides of the world, tree experts have discovered evidence of responses to pain in the form of electrical impulses within root systems.[84] The similarity to the electrical impulses within the human brain is intriguing. Others are experimenting and trying to understand how and why plants generate and respond to sound.[89] Corn kernels emitting low frequency pulses might be the base players in a band that is jamming next door, just outside of the very limited perception zone of the human ear. Ability to communicate is a human expectation of "intelligent" life forms, but from the evidence found by scientists who aren't afraid to look at the unusual in the hopes of learning something new;[12,80] humans might be the ones whose communication ability is limited.

These discoveries may just be random activities within specific plants that we don't understand. They can't be consciously perpetrated because plants don't have brains, right? Think again. Scientists have observed that the tips of roots have cellular structures very similar to animal (including human) brain cells.[43,84] The portion of each plant that we see is the product of roots that can search out nutrients in the soil, locate and absorb water, eject wastes, support the growth of a body above the ground that mirrors the mass snuggled comfortably underground, and produce energy that is passed on to other living beings through mutually beneficial partnerships. Plant roots are gracious neighbors, partnering with (non-plant) fungal species to establish extensive webs in which nutrition, energy, and chemical messages are shared between diverse species and various plants of radically different genetic makeup.[36,43] We humans have only begun to consider the depth of cooperation plants exhibit as the foundation of healthy ecosystems, but the initial evidence for potential beyond our current understanding is exciting. The communication gap between humans and plants is a hurdle we avoid when we choose to only look inwards.

What we know about the human body and how food affects it has been explored for longer and is more familiar because it is tangible and part of our daily existence. Still, the consequences of subtle variations of nutrition on the body are often perplexing. Connecting the dots between cause and effect requires equal measures of observation, intuition, and common sense.[2,21,50] While the basics of what constitutes good nutrition standards are firmly established, the study of the complex processes at work when we eat continues. This is where the general public's trust is vulnerable to manipulation by media and advertising for the sake of profit,[8,22] and the reputation of science takes on a bit of tarnish by association.

Discoveries of one chemical reaction or another, taken out of context of the larger nutrition umbrella and not given the time to be considered from interdisciplinary perspectives, are presented to the public as the New Big Thing.[8,15] It is understandable that our relatively wealthy society, facing a variety of diseases and the expenses of a broken health-care system would be desperate for a solution; we are primed and susceptible. For four generations, the country's population has increasingly been fed a diet of processed foods. The nutrients lost during processing and the chemical changes in what remains diminish the functioning of our brains to handle information.[2,16,17] Studies have even shown that the toxins we ingest build up and are passed on from one generation to the next and influence genetic development.[72,73] As a result, we may lack the perspective to counter our decline with a diet that can help us heal, and instead seek miracles. Our culture has developed a belief system that wealth can cure all by applying science, and we are willing to pay for endless gimmicks backed by pseudo-science while we complain that fresh whole foods cost too much.[15]

The nutrition industry was born of scientific discovery related to food, and both are intimately related across interdisciplinary lines to economics and politics. Marion Nestle is a nutritional biologist and well-known author who has written extensively on this topic. She clearly demonstrates how regulatory agencies and policies have been created to protect the public from the most dangerous and misleading claims of profit seekers, yet are also held back in their mission to a degree in the interest of economic growth.[8] Nutrition education is a stated government mission[65,90] that ebbs and flows in its effectiveness as a result of trying to balance between consumer well-being and wealthy corporations that influence those in power through targeted funding.[8,15]

The government is commissioned by the people to protect personal rights and promote economic development, two often competing objectives, and is held responsible if harm comes to either. How can the government be effective in either duty unless the people that make up our communities set additional guidelines and expectations? The answer is that to meet the goals of prosperity for individual and economy at the same time requires completely redefining both goals.[6] Instead of maintaining the bare minimum for healthy food standards, the focus should be to improve the standard and access for all. Instead of bowing to large corporations whose profits benefit a small number of wealthy executives, smaller locally owned companies should be promoted so that communities benefit as a whole from profits.[91,92] The action of redefining goals also needs to be applied directly to the other food science category, agriculture.

The agriculture industry is a perfect example of the way science as a field of study can become stunted and stagnant. Discoveries over the last two centuries, such as combustion engine-powered machines that could cultivate large tracts of land, developments of petroleum-based fertilizers and pesticides, and genetic manipulations of seeds are marvels of scientific discovery and exploration. Agriculture has used science to produce certain grain crops ("cereals" in the industry) in staggeringly huge amounts, resulting in the development of the calorie high/nutrition low processed foods that fill our grocery stores today to use up the excesses grown. At the same time, advertising and media have been enlisted to convince the public these products will save us.[8] A mirage has been carefully constructed for our society that time spent cooking and even eating holds us back from doing other *more important* things, and that money spent on food would be better served buying some shiny object.[6,9] The result is that we as a society are convinced that the industrial food system uses science "to help set us free," and we cannot survive without this production. This mentality conflicts with nutrition science that demonstrates a diet of whole foods helps prevent physical disease and supports emotional and mental health.

Agricultural science is also held up as a standard for progress. The wealth of our country is bolstered in no small part by the agriculture industry.[8] Every single facet within the industry supports alternative industries: farm hardware large and small, chemical additive production, petroleum production, seed and plant cultivation, raising and slaughter of livestock, food transportation, the processing and packaging industry, government and non for profit agencies that monitor and regulate, grocery and farmer's markets, restaurant and prepared food production, craft and other specialty production, home meal delivery, and waste management throughout the system. The list is endless (check out the chart in the Food, Politics, and Economics chapter for a visual on the magnitude of the larger agriculture and food industry system). Each of these businesses benefit from agriculture and science, provides business owners an income and generates tax revenue, and demonstrates the progress made in advancing our society from its simpler beginnings.

The progress celebrated has helped to elevate us as a whole, but it is not perfect. The poorest in our communities still suffer daily with food insecurity, hunger, and malnutrition as the norm.[6] Localities around the country are routinely faced with weather disasters that impact food security.[31] Social scientists Banerjee and Hysjulien note, "Beyond concerns over hunger and food safety, food disasters can disrupt both the fabric of social order and the legitimacy of political institutions."[51] The vast number of people living right on the edge with regards to food security has become even more evident during the global COVID-19 pandemic we are currently dealing with. At the same time, industrial agriculture is producing nearly twice the daily calories needed per individual; a statistic proudly noted by supporters of agricultural technology without any mention of the difference between calories and nutrition.[41,42] These calories come primarily from cereals that are the base ingredient in processed foods and have fed a health crisis in our country; one where we are concurrently starving nutritionally and battling weight gain due to excessive caloric intake.[2,75] Obesity, diabetes, and heart disease rates have exploded in the population as a result of diets high in processed foods.[2,8] For those who are food insecure, buying the least expensive highly processed foods may mean the difference between eating or not.[6] For those in the middleclass, cultural pressures create a situation where processed food becomes an attractive way to conserve money for other purchases less related to actual survival, and more related to status.

Producing the processed foods that so negatively impact our health requires agriculture on an industrial level that completely separates us from our food producing partners, the plants. Our culture approaches farming as if it is a war against nature[41,42] and employs science to develop chemical weapons to achieve what we perceive as control.[44,68,78] Science has provided evidence that various pesticides used historically in industrial farming have been linked to cancer and other diseases.[68,72,73] Regular uses of fertilizers and pesticides that harm the environment continue to be allowed,[36,44,69] and fungicides that preserve fruits and vegetables for transport achieve their results through methods we don't completely understand and may not be entirely safe.[59] Industrial farming ignores practices that maintain healthy soils in favor of mass production of a limited number of plant types, claiming progress through science is worth the cost.[36,41,42,44] The science of nutrition and the science of agriculture are at odds in the battle of progress, and all we must do to find the victims is look in a mirror.

Progress is not an evil word or concept in itself; progress can represent achievements and mile markers in our lifestyles that are uplifting. It simply comes down to what is defined as progress and why. Here we find a strong interdisciplinary connection between valuing food, politics, economics, and science. Individuals, communities, and our society as a whole all express sets of values, and those values guide how we are governed.[6,9,21,22,37,47] The most common (and arguably effective) way for the individual to demonstrate their values is through their consumer purchases,[57,91,92] provided they have the means to do so. What we actually buy (and eat) defines our values on a much deeper level than any verbal claim of ideals.[15] From the declaration of our values through purchases, the societal concept of progress is established, and science is employed to explore and identify the technologies needed to achieve that progress.[8]

Presently industrial agriculture utilizes science to develop technologies with a stated mission to improve production (of cereals mostly) using the most efficient (least costly) means possible.[41,42,77] On face value, many people are likely to align an increase in production with their own values, because it sounds as if the industry wants to produce more low-cost food to help address food security issues. Developing technologies to meet the mission of the industry is supported by the government, universities, and corporate investment as a result of the altruistic mission tone.[8] Yet if progress is to be redefined to better address both the food security and health issues impacting us in this moment, then technology also needs to be refocused.

As a society, our survival is at risk unless we are able to formulate a vision that is inclusive of both the present and the future.[49,93,94] We can potentially use science and technology to meet a restructured concept for agriculture and health (human and environmental health).[95] Efficiency and volume can take on a completely different look when combined with descriptors like nutritious, sustainable, equitable, and ethical.[33,96] Technologies discovered through scientific exploration can help us move in whatever direction we choose with few limits; they do not constrict us to the conventional industrial agricultural norms that some suggest are the only options we have to feed a growing global population. Ethical considerations like protecting the environment and making healthy whole food options available to all regardless of financial means do not have to represent a roadblock to progress; instead, they can be the guide for progress.[7,44,78]

If a hopeful, more secure, and equitable path forward is possible, why does our culture cling to industrial processes and use science without a thought to ethics? That is the question that gave birth to this book, and I am no closer to finding any kind of reasonable explanation than I was when I began researching in January of 2016. I can't fathom why scientists would engineer nanoparticles to increase the photosynthesis in plants without understanding if the natural limits are in place for a reason, if the nutrient quality and quantity suffers as a result of our interference, what those nanoparticles might do once we ingest them, and if the plant suffers as a sentient living being from the process.[12,97] I don't consider meat substitutes grown in test tubes a legitimate solution to dealing with the climate and health issues related to raising industrial livestock, especially as it requires even more petroleum-based energy than current agriculture methods.[98] While I understand and applaud the intent to help solve food security and environmental issues that inspire the scientists working on these projects, it seems like the solutions proposed only introduce new problems. These examples demonstrate that just because we can doesn't mean we should. Ethics-based science should be the standard, not the outlier. The interdisciplinary connections to ethics will be explored more deeply in the Food, Culture, and Ethics chapter to come.

To improve our food systems, we should be looking for ways to reduce the amount of contaminants that are already found in our food, not increase them.[67] The research driven by industrial agriculture to deal with water and energy issues that threatens large scale mono-crop production continues to look towards increasing more radical technology rather than acknowledging that there is a simpler solution. Scientists like Hersh *et al.* repeat the same fear-based message used by the industrial agriculture industry as a whole, "Without the addition of organic and non-organic fertilizers, the current agricultural production cannot sustain the needs of humanity."[99] While it is true that industrial agriculture cannot continue as it is due to the toll it extracts from the environment, the solution isn't bigger, more complex, and harmful technology. Smaller diverse farms using methods that protect soil health and quality can produce vast amounts of nutritious food equal or exceeding current industrial levels;[21,33,36] what is needed is to return to a partnership with the environment that supports us. Science must be employed to educate the public that there is a difference between food produced with clean or certified organic methods and those produced though conventional industrial agriculture.[70]

Scare tactics are common in the language of industry experts.[76] Statements that claim organic production cannot produce enough food to feed the world, and that small farms are economically uncompetitive are accepted as truth by the general public because of a decade's long media campaign reinforcing that mindset.[41,42] The reality is that CSA groups,[32] urban gardening groups like the one artist and urban farmer Kate Daughdrill leads,[100] and passionate food fanatics everywhere[33,34,35] are resisting this conventional statement and showing how hollow it really is. Gardens that employ permaculture[101] and polyculture[33,36] techniques can produce a wide variety of fresh foods for most of the year, not just the limited harvest season most people associate with late summer and early fall.

Can the entire agriculture industry be re-imagined overnight? No. There is much work to be done both in helping the environment to return to a level of fertility destroyed through industrial agriculture, and in educating consumers and allowing their tastes to adapt to a healthier diet.[15,50] Science, technology, and a new definition of progress need to be repurposed as the building blocks for this movement.[54] Public perception must be transitioned from the established belief system of industrial production and be inspired to participate in something better;[55,79] and nutrition science tells us that our current diet may limit our brains from being able to process that idea completely until the change in diet begins.[2,16,17]

It is human nature to resist change, even when that change inspires hope and connection. Perhaps that too is an adaptive trait of our personalities in relation to diet. Resisting change does not prevent change. Instead, resistance prompts us to ignore possible actions that may minimize negative outcomes and magnifies the probability of a traumatic event in the future. The reality is that disasters are the wakeup call we often need to drive change because of systemic denial prevalent in our culture, and unfortunately, we have been presented with a worldwide disaster in the form of the COVID-19 pandemic that will impact our daily lifestyles, our economy, our society, and our food security for many years to come. I am an optimist; I always search for the bright note in any situation. I must admit, I want to find some meaning in the (still growing) 1,010,000 plus deaths in the United States and almost six and a half million deaths worldwide due to the COVID-19 pandemic. This is a moment in time that provides an opportunity to re-examine our beliefs and goals. If we can use this event to transition to better foodways that improve our health and the health of the environment, I see it as a way of honoring the memory of those lost.

Moving forward from any disaster requires redefining what principles will anchor our vision for the future. With regards to food, defining a new focus means deciding to promote health of the individual and environment, strength of local economies, and equitable access to food by all above other objectives. As part of the shift in our cultural perspective on food, examples of local communities committed to equitable health and well-being need to be held up consistently for all to see.[38,55,56] Media leaders must position the positive aspects of food with the significance deserving of the one thing our survival depends on. A national discussion about a healthy diet and how that relates to a healthy environment must be started and perpetuated indefinitely until the public is re-educated to the reality that the status quo is a failure. Awareness needs to be raised that despite the excessive calories available to us at bargain pricing, we are nutritionally starving to death slowly. The public at large must demand that scientific study be used to identify new solutions for food insecurity, and technologies be employed with ethical standards that do not compound food security and health issues. For that to happen, the public trust in the industrial agriculture industry must dissolve and a standard of ethics needs to be reintroduced to food production.

Dispelling the doctrine that food should be considered a commodity, should be produced as cheaply as possible, and doesn't have any value beyond satisfying hunger, will bring benefit and growth to our society. Science can be an asset in helping to eliminate the disparity in food access, but local communities are the driving force in establishing through education new food values to stimulate integration into lifestyles. Beyond improving our physical diets, we need to consider the evidence within the social science field regarding the importance of food on an emotional level. How sharing meals and food fosters connection, well-being, and security.[9,21,33,37,39] We need valence, defined as the emotional quality of an idea, to enlist support and integration of healthier foodways into our communities.[46]

Science connects in yet another interdisciplinary way with the social sciences to help form a well-rounded picture of where we have come from, where we are today, and the potential outcomes for our future. Our society as a whole is trending toward a more fragile self-confidence.[6] I often wonder if this is related to our strained relationship with our food sources. Case studies about communities uniting around local food sources demonstrate to me that besides the health benefit, these groups of people are interested in a deeper connection to each other. Connection with food sources is often not an explicit goal, yet it is a significant tool that draws the community together.[38,47,55,56,79,102] Exploring the psychology of how choosing fresh foods impacts the self-esteem presents a chicken or the egg scenario. If someone makes the effort to choose whole foods, they are demonstrating an interest in their own well-being.[6] That choice repeated regularly contributes positively to their well-being, physically and emotionally. The habit developed supports and sustains itself and the person's improved health increases their overall confidence and self-esteem. We have unending examples of the opposite behavior within our society; it makes sense that the positive incarnation would be equally effective.[46]

Here again, we see an interdisciplinary connection to those who have studied the relationship between food, religion, and spirituality. In industrial agriculture, livestock raised on feed sources that are not natural and housed in completely abnormal and barren surroundings has resulted in meat that is full of stress toxins.[37,58] The shift of a significant number of people in our society to vegetarian and vegan diets is linked to a moral and ethical struggle as much as a physical rejection of a food source that makes us feel sick.[103] What else may we gain from maintaining a more direct and respectful relationship with our food sources? Social science studies show that the power of positive thinking and actions promotes a healthy perspective and allows us to enjoy experiences, rather than rushing through them.[16,17,50] Promoting positive thinking increases the memories that give meaning and value to our lives.[6,39] Combined with a diet of fresh whole foods regularly shared with those we care about, the pleasure of living thoughtfully in this moment and looking toward a bright future is a perfect integration of the physical body's chemistry and the intangible spirit or soul.

Acknowledging the link between our physical body and the essence of our personality can provide prospective for new questions to be explored using science. What if there is some invisible energy passed between two living beings when one is consumed?[29,45] This idea is discussed more in depth in the chapters about health, and religion and spirituality, but how can science help us address this question? We are only beginning to recognize that plants have energy fields that react to stimulation.[12,80,84,88,104] There seems to be enough circumstantial evidence available through simple observation, as well as traditional Indigenous teachings, to at least consider beginning a debate on how to measure whether energy transferred between food and eater is more than just caloric.[13]

The topic of connecting and communicating with our food sources might seem less unusual to those with some gardening experience. One doesn't put forth the time and effort to maintain a garden without getting something back from it. Visual stimulation is only one piece; gardeners are rewarded on a much deeper level by the joy ones feels when connecting with and nurturing another living being. Growing any food plants, even a window box of herbs heightens that experience further because the reward becomes part of the gardener. Only a few of the people I know have a vegetable garden annually, but they all agree the experience is worth it. The act of caring for a food plant that spends an entire growing season to give us a tangy red tomato or a crunchy green leaf, the burst of juice from a peach, strawberry, or blueberry that just finished basking in the sun; nurturing contact and patience is rewarded in flavor, life-enhancing nutrients, and something more. Building scientific studies that increase understanding of this connection would help raise awareness by the public and support the shift in vision for the future discussed earlier.

How could someone know a level of intimacy if they have never been close to the plants that support them? Without knowledge and experience, it makes sense that the general public wouldn't be passionately pushing for science to step up find out the details behind the ethereal connection we have with our food sources. To truly value the fresh food we eat, one must understand where it comes from naturally. I have heard and read several stories that poke fun at children and even adults who don't know what part of the plant carrots or potatoes are. When I led the food valuing exercise outlined in Chapter 2 with an apple in a room full of people, a couple of participants expressed that they had never considered that the flower is the beginning of the formation of the apple. I don't think it is conceivable that everyone should begin growing all of their own food; however, learning to nurture even one food plant can help us develop a connection to our food and where it comes from. The exercise also helps promote respect for those in our community who dedicate their livelihoods to growing food and aids us in better accessing valuation in relation to nutrition when budgeting for groceries. These changes in attitude will determine how science is used in the future with regards to food production.

Eating more fresh whole foods is important, but I am certainly not the first person to say that. I think the reason the message is not taken to heart and real change isn't triggered, is because of our ambivalence toward recognizing the living beings that are the origin of food. Changing that perception will inspire and generate support for scientific study that isn't controlled by corporations with profit as their only focus.[13,105,106] The model of big business is to continually grow more product faster,[91] but big business doesn't have to be the accepted model for our food. Philosopher Baldwin says is perfectly, "Choosing food grown in more lively, rather than less lively communities is a significant move, and constitutes a practically applicable ethic."[22] Science can support our hunger for a new understanding and processes that allow for fresh food, sourced locally and seasonally to improve food security, food equity, and nationwide health.

CHAPTER 5

FOOD, POLITICS, AND ECONOMICS

As I write this chapter, I am acutely aware that much of our nation is weary and frustrated by the politics of the day, regardless of whatever position each person identifies with. I sympathize with the impulse to dismiss the importance of considering how our food systems are influenced by local and national government policies, and how food and money are related because the topic is unendingly complex. I beg you to be patient and stick with me here while I explain what I have discovered through comprehensive research: that despite the bureaucracy, as the consumer we do have power and we can prompt change more easily than is often thought possible.

Overall, what I have discovered in my investigation is that politics and economics in relation to food are not institutions divorced from everyday life. It isn't "over there" in the corner and able to be dismissed or ignored. It governs us and is governed by us.[8] It is designed to protect us; those we elect to handle the day-to-day operations need to be held accountable to ensure that it does. The food system is a necessity; we must eat to live. More importantly, what we eat helps determine how we live.[2,21,22,37,102] All the stakeholders in the food system play a part in determining what is available to the consumer to satisfy their demands. To have better options, we must demand better options based on a value system that promotes health for the individual, their community, and the environment. Technology is an important tool and should be utilized to improve the food system, but technology must be guided by the ethics and values of the educated consumer who can competently understand that we need to protect the resources that feed us.[36,44] Technology cannot be promoted simply for its own sake, for the profit of a select few, or because of misinformed fear without careful consideration of the implications it may bring.

To begin to better understand this immense topic, like in other chapters, it will help if we are all on the same page with some basic terms. According to the dictionary "politics" is described as the science, art, practice, or profession of conducting government affairs.[107] While everyone likely has an understanding of how their life is impacted by politics to some degree, my impression from the many conversations I have had is that most people do not regularly take a step back and consider the extended repercussions of making changes to our governmental process. I think the one thing most people can agree on is that change is needed, even if the how is hotly contested. No one person has the perfect solution for improving the politics of our food system, we must work together to build ideas into concepts and concepts into practices. The suggestions I make in this chapter are intended to trigger discussion and vetting; those discussions may give birth to what it will take to redesign and refine the system into something that works for more of our society.

Less understood is the term "economics," defined as the science that deals with the production, distribution, and consumption of goods and services, or the material welfare of humankind.[108] So many people in today's world seem to perceive economics or the economy as related directly and completely to the stock market. The stock market is simply a speculative investment tool that is reactionary to what may happen in the actual economy. As part of this chapter, I will do my best to demonstrate the variety of connections that make up the food economy and the intricate ways they interact. Demystifying the food system economy is crucial for individuals to understand that they are an autonomous participant in the system rather than reliant on the whims of others. Now that we all have some clarity on the meanings of the subject matter in this chapter, we can explore how the food system relates directly and in interdisciplinary ways to our daily choices.

Living in a free society is a basic founding principal in the United States, making it hard to reconcile the variety of influencers who exert control over what we eat while the illusion of consumer choice is carefully maintained.[6,8,21,44] We are inclined to feel confident that our food simply comes from farmers, while there is a vast network of industries that stand between farmers and consumers. Most of these industries have been developed with a purpose of adding value to improve the consumer's experience and in principle they do, yet in many cases the net value of the product is reduced overall during the process. It is this exchange of value versus addition of value that is often misrepresented to or misunderstood by most. To fully appreciate the value, comfort, and convenience of the products that improve daily life, one must recognize that the resources they are developed from are limited. The Earth is mined for metals, chemicals, water, and fuel to create machinery, plastic, glass, cardboard, and food additives. Countless varieties of living beings (plant and animal) are enlisted in servitude and generously make the ultimate sacrifice once they mature so we may eat. All of this is necessary for the most basic of food production and delivery. To ignore these elemental components of the food system and drift from meal to meal assuming that some far-off entity is carefully considering myriad factors to ensure health and sustainability for the individual's benefit is dangerous. To comprehend the inherent risk in the food system, we must recognize what that system encompasses.

The food system economy is massive. Food is the one thing everyone must have; it does not last and must be replenished regularly. Personal perspective and financial position determine which parts of the food economy are acknowledged and valued by the individual. Understanding only part of the system can set the stage for break downs that go unnoticed until a larger failure of the whole is imminent. Think of it like a car; regular oil changes are needed to keep the motor functioning properly, yet if the oil is forgotten about the car will not stop running immediately. Over time, the motor will start to get weaker until at last when it has been damaged past the point of repair, the car will cease to run. Unlike a car which would need expensive repairs or to be replaced, a food system collapse would result in a lack of food for countless communities, jeopardizing human lives. In this scenario, it would not matter if the consumer had financial means as there would be nothing to buy. Belief that this plot is possible only in books or movies, not in real life, demonstrates alarming complacency. We should not be panicked about the food system, but we must be cognizant that it is a complex system that is always changing, and importantly, that each person contributes, takes from, and depends on this system regardless of their financial status.[6] As such, we must all be alert and proactive regarding opportunities that strengthen the food system and recognize current hints of malady for what they are before they trigger decline.

Symptoms that the health of the food system economy is in jeopardy and is being maintained on life support are clear. Small family farms are disappearing because they can't compete.[44] Large agribusiness farms are given subsidies (money from the government) to manipulate what and how much is grown.[8,41,44] Chemical fertilizer and pesticide makers, food additive, and food processing companies are required by stock holders to continuously grow profit margins, forcing the manipulation of food throughout every step of the system rather than using technology and science to limit manipulation and provide the consumer with food that has not been tainted and polluted.[8,21,41,44,91,92] Across the food system, the general fostering and reaffirming of the consensus that imitation and substitute food is nutritionally equal to whole food, and necessary in order to feed the nation[41,42] not only provides a false sense of security for the consumer, but also dissuades existing companies from striving to improve quality.[6,8,21,44] At the same time, it discourages new businesses from seeking to find alternatives that promote purity and partnership with the environment that supports us, forcing them to chase profits over perpetuating integrity.

Acknowledging dysfunction in the food system means discerning how various stakeholders in the system interact, and how they are influenced by others. With a goal of eventually visualizing a new economic food system supported by government policies that improve the health of the consumer, it is critical to understand the existing system and where it falls short. For me, a visual aid was necessary to really grasp the complexity of the current food system in the United States. As I pieced together how various industries impacted others, I created this chart: *See next page*

FOOD SYSTEM ECONOMY CHART LEGEND

Petroleum: Used to power everything (directly and indirectly) from the machines that develop resources, farm equipment, transportation vehicles for food, labor, and delivery, to the appliances that aid in food preparation and storage. It is also a raw material from which fertilizers, pesticides, food packaging, and food additives are made.

Resources: The raw materials to produce these resources come from the Earth. They are then manufactured into hardware and machinery, chemical fertilizers and pesticides, and seeds.

Farmers: There are many types of farmers; most fit into one or more of the following categories: large conventional "agribusinesses," large specialty (including organic), small specialty (including organic), livestock, and technology based.

Political and health entities including the government (local and federal), regulatory agencies, lobby groups, consumer watchdog groups, municipal water resource managers, and health professionals and researchers as their mission is to preserve the health of the country.

Logistics, Waste management, and Recycling: Once food products are harvested, they travel via domestic transportation, as well as import and export companies. Food distribution companies help to organize the process, NGOs (nongovernmental organizations) work to ensure those with limited resources are provided for, and the waste management and recycling industry ensures that communities do not become overwhelmed with unused products.

Labor. The food system economy as a whole supports an extensive labor force. Most employment numbers are calculated for the individual industries that make up the food system rather than for the entire system.

Processing and Marketing Industries: Food packaging production, food additive, and food processing companies. Included in this group is the massive advertising industry who claims to add value to our lives via a constant stream of marketing, so we are aware of what is available to meet our daily needs.

Retail outlets: Grocery stores, the restaurant industry, wineries, breweries, and distillers, farmer's markets, CSAs, and convenience and vending services.

Food: All of the food available in the system, including whole foods, substitute and imitation foods.

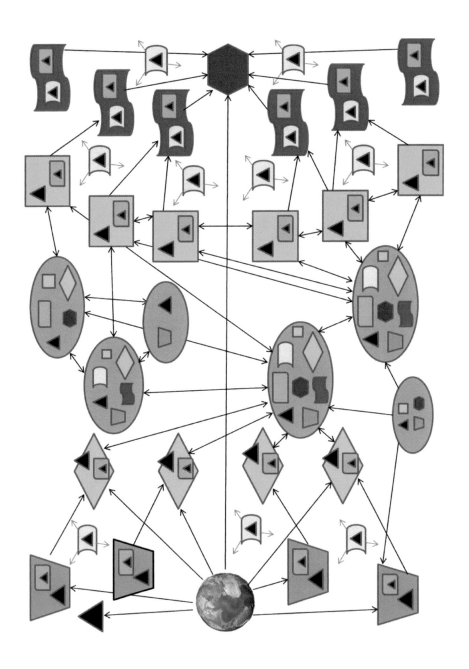

The complexity of the food system economy becomes really clear in this visual representation. The various values that are added and those that degrade the food system as a whole really stunned me. I had considered many of the relationships individually, but the magnitude of the entire system is incredible.

 Petroleum

Perhaps the most surprising aspect is that the petroleum industry plays a role in absolutely every step of the process.[21,49] It was a wakeup call for me, and I began looking at how I could reduce the amount of petroleum I use even by a few percentage points to minimize the negatives related to this resource.[109,110] If each consumer committed to a small reduction like this, the positive impacts would multiply exponentially and perhaps buy us time to find alternate ways to power our lives before we are forced to deal with the exhaustion of this resource.[49,109] In addition, the food system would become healthier and more resilient. Sustainable. My other revelation was to look at each and every media source with new eyes; to think about the language used in relation to food. Nutritional biologist and author Nestle describes countless ways how the message "eat more" is pushed on consumers by both advertising agencies in an effort to sell products, and government agencies whose conflicting mandate is to provide nutrition advice, encourage business growth, and regulate the advertising industry.[8,15] We expect this guidance to be impartial, but instead, it is weighted toward pacifying businesses that spend fantastic amounts of money lobbying to influence the system.

 Resources

Looking at the complexity of the food system is like visiting a new place for the first time. The lights, sounds, smells, and movement of people are all jumbled. Your brain jumps from one sense to another trying to process and organize information, instinctually evaluating for danger and opportunity. Let's take a step back and look at each layer that plays a part in the process of providing food to our population. We might expect to see farmers as the starting point for food, but farmers cannot grow without equipment, energy, and seeds. The number one provider of resources needed to grow food comes from the Earth. Mining and manufacturing industries work to produce machinery and hardware ranging from massive to miniscule to fit every application, and the petroleum industry seeks to fuel the entire fleet. In just this first step, we can find several opportunities for the economic development of recycling and alternative energy industries to reduce waste and improve efficiency. As previously noted, petroleum is a finite resource; as it becomes harder to find and extract, the cost of food will be negatively impacted and become more expensive. Developing alternative energy sources will require existing machinery and hardware to be upgraded. I think it's obvious that we should begin this process while we still have the current system to rely on rather than wait until food becomes scarce to get creative.

Next in the system are the industries that collect the progeny of plant beings who absorb sunlight, water, and minerals and use them to grow the living tissue that is our food. Types of seed companies are diverse and use contrasting methods, from the most natural practice of cross pollinating for plant evolution to gene manipulation, which combines traits from dissimilar species to attain a specific quality. Debates abound regarding methods used, but the seed industry is necessary in whatever capacity the public decides to allow it to flourish.

In conjunction with seed companies, the chemical input industry produces pesticides and fertilizers. Developed to support the growth of only the seeds farmers' plant, these chemicals promise destruction of any living being that interferes in the farming process. This relatively new industry (only a century or two old) born of technology and science has found ways to increase efficiency and production but has often neglected to self-evaluate or outright ignored the secondary consequences, that result from their concoctions.[68] They choose to use their resources to establish a belief within the public that they are essential, versus challenge themselves to return to technological discovery to cure the problems their products have caused.[8] A testament to their effort is that much of the public has embraced the mindset that there is no other option and forgotten all that we have learned about soil health since the first human planted a garden instead of foraging.[36,44] Here, there is plenty of potential for the food system economy to both evolve and grow, the only question is whether to follow the path we are on or choose a new and cleaner one that doesn't hurt the environment and instead works with it.

Farmers

Finally, we come to the farmers. The term farmer is a very broad descriptor for a multitude of people who grow food. Farmers and their farms can be big, small, rural, urban, organic, conventional, technology savvy, technology innovators, specialty, staple crops, polycultures, livestock, hydroponic, grain producers, local, backyard, and CSA; the descriptors are endless. The current economic food system tends to pit farmers against one another in competition.[44] Farmers can fall prey to the same cultural obsession with profit as the rest of the nation, faced with choices between producing as cheaply as possible or choosing to invest their time and effort in cultivating a quality product that will be desired for its superiority.[23]

One might assume that a farmer's income is directly related to market prices, but the amount of money spent on food that actually reaches many farmers is pennies on the dollar.[8,21,23,36,37,44] When food prices go up, rarely does the farmer see in increase in income. Larger farms rely on the government to supplement their income if they qualify (most small local farmers do not benefit from any subsidies[8,36,44]). Smaller farms must decide between selling to retailers or developing a local market to reduce the number of companies who take a portion of their income.[21,23,32] Both the farmer and the consumer suffer for each of the intervening stops along the chain from farmer to dinner table; the farmer receiving a smaller fraction of the food dollar spent, and the consumer who pays inflated prices for more travel weary food. With farmers focused on survival of their livelihood, and consumers who are interested in paying the least amount for food as possible, it isn't surprising that protecting the health of the environment gets pushed off as a secondary priority even though we cannot continue to grow food without it.[21,32,36,37,44,101]

 ## Political and Health Entities

The next entity in the food system is the government, whose influence can be observed in several ways. Multiple agencies such as the USDA (United States Department of Agriculture), the FDA (Food and Drug Administration), CDC (Centers for Disease Control and Prevention), and the FTC (Federal Trade Commission), each with their own mandate to promote and/or regulate food sales are involved in the growing, transporting, slaughtering, processing, advertising, and retailing of food.[8] These agencies do not handle the work of getting food ready for purchase, but they set the standards for how that work should be done and monitor to ensure the rules are met. Some of these agencies also test and review evidence regarding chemicals, additives, and food packaging, to determine the general short-term safety of ingredients that make their way into our foods. The power these agencies have, and the missions they are tasked to handle are determined by legislators who are elected by each of us and funded with our tax dollars.

Lawmakers manage the high-level functions of these agencies with varying degrees of independence and set the budgets for them to fulfill their duties with. They manipulate the market and encourage targeted production using subsidies, tariffs, and check offs, for the purpose of creating stockpiles that can be held or traded.[8,44] These actions are touted as promoting free market economic growth in the food system; however, the consequence is an artificially balanced scheme that costs the consumer and stifles innovative development. Leaders in top positions also negotiate on the global stage with agricultural products to wield power.[6,8,30,44] In the hands of the government, food is reduced exclusively to a commodity;

it loses its vital quality as a life-giving and life-sustaining relationship between species.

At the local and state level public officials help to negotiate whose rights take precedence over another with regards to the environment, often with a preference to those businesses that help support their political aspirations with financial contributions.[6,8,44,109] The environment, clean water, healthy soil, and strong food production may not seem connected to campaign finance reform at first glance, but the only way to ensure our representatives are working for their communities is to eliminate the distractions that force them to choose between professional survival and the community they swore an oath to serve. We elect representatives to government positions with the expectation that they will ensure the consistent delivery of healthy food to every community, but it is our responsibility to monitor their effectiveness in this role.

Yet another way that the government is charged with protecting consumers is to require clear product labeling and advertising.[8,21] Agencies like the FTC (Federal Trade Commission), the FCC (Federal Communications Commission), and the FDA (Food and Drug Administration) all share parts of the responsibility. Perhaps this is the area that needs the most work and reform when it comes to government regulation and food. Labels and commercials are consumed by a public who trusts the information they receive is accurate. Nestle describes at length how some companies post all manner of claims that are quite clearly meant to distract the consumer from the negatives of the product, while others have to fight quite vigorously to be able to declare their product absent of certain ingredients.[8,15] Organic certification and non-GMO labeling are just two examples of this unbalanced leverage. Chemical companies know that if given the choice, the consumer will pick a product that does not have chemical residues. Rather than develop products that are more natural, they petition the government to limit organic labeling and deliver studies focused on the "safety" of food grown using conventional methods.[8,21,37] Oddly, the money spent to conduct these studies, which is in the billions, is more than the anticipated loss in profits if consumers decide to choose more foods grown using organic methods.[8] Discovering what the objective for spending so much money is could help shed light on what really matters to these companies.

Conventional thinking suggests that the higher cost of organic options is somehow related to the growing process, but the truth is that organic products and their prices include the overhead of certification, while prices for conventional produce are artificially controlled using tax-payer money.[8,21,37,41,44] Organic produce is the fastest growing segment of the agricultural economy at this time.[2,8,54] Investing in small organic farmers to increase certification would be a wise investment versus actively trying to limit the amount produced. Organic vegetables do contain less chemical residue in their tissue,[70,71] as discussed in the health chapter. Because the government is partially responsible to protect the health of consumers, one of the interdisciplinary connections noted several times in this book, to me it seems obvious; instead of providing subsidies to farms using conventional industrial methods to limit production of staple grains, taxpayer money should provide an inexpensive pathway for organic farmers to demonstrate that they meet and maintain strictly set standards.

More recently the fight against non-GMO (Genetically Modified Organism) labeling has also grown.[8,21,37] The food and health chapter of this book addresses the debate regarding the safety of GMO foods in more detail, but the political and economic perspectives can't be ignored. Some consumers have a strong emotional reaction to GMO foods; labeling foods that purposely don't contain GMOs should be a common sense decision to provide the consumer information they want, yet companies who produce GMO seeds and those who grow them have fought very hard to prevent such labeling.[8,21] Allowing a label suggests to the consumer that there is a difference, and it might influence their choice not to buy a product that contains GMOs.[8,76] Rather than seeking to better understand the consumer's concerns and focus technological development in a different direction, these companies pressure the government to prevent the consumer from making an informed decision by fighting labeling criteria. The most common argument used is that GMOs are necessary to feed the growing population, yet subsidies continue to be paid to farmers to limit how much they grow to avoid surpluses.[41,44] This irony is also evident in the regular noting in both cited and un-cited literature that our farmers produce more calories per person than are needed for a healthy diet while the obesity crisis grows.[2,6,8,21,36,37,41,65,91] Regulations abound to support and protect certain portions of the agricultural sector; threatening that without them food security is at risk, while at the same time, they brag about metrics that ultimately have nothing to do with health and well-being.

As an alternative, manipulation of the food system by government policies should arguably be aimed at ensuring healthy food is accessible to all, and that consumers are educated accurately about what is healthy. Assistance programs are important to deal with hunger and food insecurity in the short term, but long-term plans to provide local governments with the resources to encourage development of more stable and balanced food sources in both urban and rural locations are crucial, yet absent.[6,8,54,102] While large agribusiness farms grow millions of acres of grains which contain the caloric count needed to meet the population's daily requirements, a diet consisting primarily of calories from grains is unbalanced and unhealthy.[2,6,22,37,61,65] Shifting financial supports to reward communities that maintain a degree of food system self-sufficiency and encourage others to develop food independence would reduce the need for support programs over time by making food more accessible and less expensive.[6,47,93] All of these different actions currently taken by the government show that the politics of our food systems operate with a short term emergency management outlook exclusively, trying to plug the holes in the bucket as they appear rather than encouraging communities to each take a cupful from the bucket to reduce spillage and allow for the bucket to be repaired or retired.

I have mentioned several times in the last few paragraphs, food processing companies resist labeling reform. The primary way they do this is through lobbying groups,[8] the next layer in the food system chart. I will admit that this is the part of the food system that I find the most disturbing. The lobbying mechanism blatantly unbalances the system in favor of those with power and resources. The elected officials we choose are supposed to "serve" the communities they come from, the entire community, not just the largest businesses with the power of the dollar. Allowing representatives to be courted to make decisions that are not considerate of the entire community is something we must change by demanding more from those who work for us. Honestly, I do not have a problem with successful businesses if they are rooted in ethics and transparency. But power is hard to resist, so there should be a mechanism to limit those who may be tempted from banding together and creating lobby groups that compound their resources into a seemingly unstoppable force. There is no necessity for lobbyists. Without lobby groups, the government will not cease to work; it will simply have to pay more attention to the communities they serve.

Changing the regulations that allow for lobbying is not a first step in the process though; to achieve the elimination of lobbyists, consumers need to be far more educated about the rest of their food chain, and how to make adjustments that directly impact them in the immediate future. Luckily educating the consumer doesn't need to be a heavy lift. Health professionals and researchers are the next layer in this section of the food system, and they fulfill roles from face to face interaction with the consumer, working for food processing companies developing products and improving others, and working for government agencies charged with food safety monitoring.[6,8] As part of this layer, there is already a workforce of registered dieticians available in many communities with the knowledge to help families develop better eating plans.[2,8] Their advice is simply muffled by the cacophony of noise produced by advertisers who provide the bulk of nutrition information in most people's daily life.[6,8] The solution doesn't require eliminating the advertising industry; it just involves ensuring that messages related to food are guided by health, science, and ethics rather than by who will pay the highest dollar.

One of the moments that caused me to pause when researching this book was that hundreds of billions of dollars—a number so large that I cannot actually or even realistically relate to it—is spent by several different food processing companies annually for advertising.[8] At the same time, nearly half of our population suffers from obesity, food insecurity, and lack of financial resources to improve their diets.[2,6,8,61,102] Being fed a constant stream of marketing messaging that "adding value" via processing to something is more than just handy, it is absolutely crucial, has an impact on perspective and preference. This is why advertising exists; because people's wants can be manipulated. Promoting a product or idea as desirable will influence people to want it regardless of its real value, and companies are willing to pay for that advantage. The very language is deceiving. While convenience, preparation time, and shelf life are routinely improved, no mention is made of what is sacrificed. Nutrients are lost, connection to family and nature is severed, and a perspective of gratitude is disassociated from the act of consuming the energy that gives us life.

I choose not to reinforce the message that the processing of food items to "add value" is altruistic, and that the benefits compound with the level of processing. Some processing is valuable, such as when I make sauce or soups and preserve them for later use. I value the processing by the farmer so that the traceable chicken I buy is ready to cook, and I don't have to harvest and butcher a live chicken myself for dinner. I am grateful for the cheese makers, tea leaf blenders, and the beekeepers that harvest honey for my tea. Still, even these artisans minimally process their products with care for a specific purpose. Profit. The determining factor between processors is HOW they process their product. Do they tend toward the ethical, aiming to preserve the consumer's connection with the whole foods being processed, or do they seek to disguise and infuse their product with inexpensive materials without concern or belief that the product will have a lasting impact on the consumer? This is what each consumer needs to consider when deciding where to spend their food dollars.

Profit doesn't have to be a negative; an economy is defined as the buying and selling of goods, and the seller who processes their product in a specific way to entice buyers deserves to be paid for their skill. The issue is when the consumer is seduced into believing that excessive levels of processing are necessary for their enjoyment, and the negative impacts of that processing are actively hidden, eliminating the consumer's ability to make informed decisions about the products they buy. To encourage a more realistic perspective, I will use the term "manipulated for profit" in place of the industry's favored "value added" going forward, with an explicit understanding that profit is not the enemy. We simply need to acknowledge all the advantages and disadvantages associated with processing in a balanced way as educated consumers.

Many government policies effectively support manipulated for profit messaging for the benefit of the economy rather than seek to protect the nation's constituents, as if the two must be opposing ideologies. If just $1 from each thousand spent on advertising was used to provide the communities that are food insecure with fresh whole foods and resources to learn how to use those foods to achieve better health, the result would be incredible. For every five hundred billion spent, five hundred million dollars would be available to improve the lives of struggling consumers. The problem is not a lack of money; the problem is how that money is used and taxed.

Depending on the political perspectives of each reader, the amount of governmental action that is appropriate with regard to advertising can be debated, but what cannot be debated is that government policies are already engaged in every single aspect of our food system. Shifting engagement from one action to another to eventually minimize that engagement in the long term while improving the health of the nation has got to be better than maintaining the status quo. The system is broken yet accepted as if there is no alternative. The public at large is frustrated with their representation, so they walk away and let corporations control the policies and messaging regarding our food systems. The taxpayer is charged for food system–related costs such as subsidies to farmers,[36,41,44] health-care issues that could be eliminated with better nutrition,[2,6,37] and inflated food costs in the retail market.[8,21,102] In the end, the corporations take home a profit, the small farm disappears, and we die from slowly eroding health.

I have spent a great deal of time on how the work of politicians, lobby groups, and health professionals currently influence food choices, as well as how they might change for the better in the future. This is no accident. Our food choices are manipulated heavily in carefully crafted ways; facing that hurdle means admitting that as the consumer we have been seduced, and we have paid the price with our health and our food dollars. We are only halfway through the list of industries that stand between the source of food and our dinner table, but the rest of the intervening parties tend to be more sensitive to the consumer's demands. This means that they will adapt to change faster than the government if improvement is demanded, and this is the best way to get lawmakers attention between elections.

Moving along in the food system we find some balance. Advocating for shoppers with government officials to combat the mission of lobbyists, consumer watchdog groups play a part in monitoring food safety ensuring that the regulations that are in place are followed.[8] Also somewhat removed from the day to day food considerations of many is the work done by NGOs (nongovernmental organizations) that operate in a not-for-profit structure to identify how surplus food staples can be delivered to those who need them most.[6,8] NGOs operate both within the country's borders and beyond with a mindset that all life is precious and valuable, and food should be a basic resource everyone should have access to regardless of social status. NGOs need assistance in the physical delivery of food products which brings us to the next industry in the web.

Logistics, Waste Management, and Recycling

Transportation companies have a hands-on role in the food economy. Moving food directly from farms to distribution centers then to retailers, or to food processing companies then to retailers; food shipments follow an intricate and almost invisible web[37] that delivers with astounding regularity. Our reliance on this industry without proper consideration could be viewed as ungrateful, but it is also dangerous as we rarely consider how we will get food if there is a natural disaster until the situation is upon us.[31,51] No corner of our country is safe, hurricanes, extreme wild fires, flooding events, ice storms, earthquakes; with little notice local communities can be isolated from the far-away places where our food is grown, impacting food security.

 Labor

In addition, we are witnessing today how a global pandemic can reduce labor forces at all points in the food system and trigger widespread unemployment numbers, causing food insecurity in communities that had not been considered at risk before. Still where there is food to move, the transportation industry does its best to move it, and petroleum is used in every single step of the process.

Import and export companies fit into the system beside transportation companies and are subject to the same weaknesses with relation to extreme weather events and global pandemic factors. They also must deal with financial penalties levied by legislators in the form of tariffs, which ultimately are passed on to the consumer in inflated prices.[8,44] Next, the food packaging, food additive, and food processing companies all touch various crops in the system with the intent on manipulating for profit by reducing preparation or increasing shelf life. As noted previously, what is lost in nutritional value is rarely noted in comparison to the gains boasted about in advertising; the consumer is left to decide if convenience takes priority over nutrition, taste, and enjoyment, without being presented with all the facts.[8,15,22,37]

 Processing and Marketing Industries

Some in the food processing industry do provide value while manipulating for profit, walking a tight rope between public service and predatory profit seeking. I appreciate that I don't have to churn my own butter, but are aisles full of unhealthy foods really providing a benefit to the shopper? Should more snack foods be developed just because they will sell? Setting an expectation of a minimum degree of ethics in food regulations for these food processors and preparers could allow for them to survive as a valued part of the economy as employers, providers of healthy and convenience food items, and as business owners who enjoy a comfortable life as a result of the hard work it takes to build a successful business. The reality is that many corporations are run by people who are caught up in the same emotional cycle of earn and spend to achieve some unattainable benchmark as the consumer; they have bought into the competitive messaging that victimizes us all. Food processing companies are not the enemy, but we need to clearly declare that we expect a higher minimum level of quality from them, not a continuously diminishing one.

 ## Retail Outlets

At this point in the food system, all the various whole, substitute, and imitation food products come together and become available to consumers through a variety of retailers from different industries. Perhaps the most traditional and common source is grocery store chains that display a dizzying array of items. A huge share of food dollars is also spent in the restaurant industry with individual locations that cater to every level of taste, and compliment it with corresponding degrees of service, from a help yourself set up to luxurious and expert indulgence. Restaurants providing full service at any level combine two different manipulated for profit features; the convenience of a variety of meals prepared on demand, and the experience of a meal often shared with companions. Offering a dining experience that includes conversation, relaxation, and nourishment is a return to an attitude of appreciation as discussed in the values chapter earlier in this book. As noted then, the meal experience is one to relish and strive for in that it promotes respect for the food that keeps us alive and allows us to take the most from the interaction. It reminds us that meals are more than just taking in calories.

While it is important that restaurants are providing this experience that has faded in our day-to-day home life, I think we need to be cautious that we don't begin to believe that they are the only ones who can provide it. The act of cooking and interacting directly with food during the preparation of a meal is important to our attitude of valuation, and all individuals should be encouraged at some point to develop that appreciation. The key is to find balance between preparing meals with whole foods in the home, enjoying the excitement and freshness of meals served in restaurants, and having previously prepared or quick to prepare meals at the ready for moments when time runs out. (In my mind, a previously prepared meal is soup, chili, lasagna, or some other dish I have made and either canned or frozen, not a highly processed meal substitute.)

Today, amid the global COVID-19 pandemic, we can observe that the proportion between meals cooked in the home and meals eaten in restaurants has shifted significantly. Having worked in the service industry for decades, and being a huge fan of exceptional eateries, it is hard to watch businesses suffer, but overall as a society, the reintegration of meals eaten in the home is important and necessary. We struggle and cling to the way things were when what we need to do is use this pinch point to innovate and refocus. Talented chefs and cooks should be looking to market their skills in ways that assist those preparing meals in the home. Takeout meals should become meal kits or classes. There are an unlimited number of ways to promote and enjoy food if we stop clutching at what was and look forward toward something new and exciting. Ultimately, I think diners will develop a sense of admiration and respect for those who provide an exceptional eating experience outside the home if most meals are freshly prepared within the home, rather than take them for granted as was becoming a trend prior to the pandemic.

Insensitivity and disregard for what it costs to provide a valuable dining out event has heightened competition in restaurants with some increasing plate sizes well beyond what is healthy or using cheaper and cheaper ingredients to keep prices within consumer tolerance. As restaurants look to find their unique signature feature, a new niche industry has broken off and expanded from the restaurant trade. Alcohol makers were once dependent on restaurants to offer their products as a manipulated for profit specialty to their menu. Now wineries, breweries, and distillers are seeking to bring the consumer directly to their locations and offering trending food items to complement their products in a complete reversal of the norm that used to draw the consumer into the public for meals. Additionally, as meals become increasingly unconventional, convenience and vending industries also continue to look for opportunities to participate in bringing food to the consumer versus forcing the consumer to come to them. Importantly, don't forget that petroleum is the silent partner underlying the preparation and delivery of food in every single one of these outlets that provide food, as well as in your own home.

Meanwhile, some farms have decided that supplying an increasing number of retailers is simply too costly for their bottom line, and they have moved into the business of becoming their own retailer.[23] This expansion allows them to develop a reputation and be recognized by the public for quality and freshness that is lost in the larger grocery setting. Farmers markets, farm stands, and CSAs short cut the system to provide the freshest products possible to the consumer.[32] Those who choose to buy direct from the farmer reap the greatest reward in nutrition,[2,8,15,21,22,37,70,71] nurture the strongest reverence for the living beings that sacrifice themselves to support us humans, and minimize the amount of petroleum used to acquire their food. This practice has become a movement referred to as "locavore" where values associated with eating expand well beyond the instinctual reaction to eat when hungry.[21,36,37]

A primary goal within the locavore movement is to ensure farmers are paid a livable wage for their work, and to strengthen community

networks. As economist Shuman, activist McKibben, and others eloquently demonstrate, supporting community-based businesses is crucial to
a strong economy [55,56,57,91,92,93,111]. McKibben notes "'Local' steps far enough
outside current conventional economics to represent a real challenge (to
industrial agriculture)."[92] Allowing larger chain stores to monopolize an
area removes currency from the locale, eventually draining it of financial
resources. With relation to food, supporting food artisans, be it the farmers who specialize in produce, livestock, dairy, or honey, or the secondary
business owners like bakers, butchers, and cheese makers, this conglomerate is a community's lifeline in emergency weather events when outside
support cannot be accessed.[31,51]

Climate change emergencies also suggest we should look to nature
as a guide and encourage redundancy for strength when building up local
food systems.[68] A village that only focuses on maintaining one local farmer's livelihood, utilizing the expansive food system for the rest of their
needs, runs the risk if their trophy farmer has a crop failure or a personal
health issue. A solid locavore neighborhood will encourage and fund multiple farms so those farmers can then find the niche that best suits them.
In an interview with a local farmer in my area, I once asked if he had considered expanding in a certain way, and he replied that there were others
already supplying that product.[111] He said that rather than competing and
reducing the income of other local farms, he was more interested in finding
ways the community was not being served and creating products to meet
that need. In that way, he might be able to funnel more business toward
all the farms by raising interest and awareness. This is the wholesome and
protective mentality that makes for a strong economy.

This is where consumer education and awareness once again become
the spotlight. As the entire food system comes into focus, it is easier to
understand the links between consumer values, governmental action, messaging back toward the consumer through advertising, and the durability
of the economy. To establish goals for the food system economy and convey those goals as expectations to representatives in the government, the

individual must understand how values need to guide the process. Allowing cultural norms to reinforce that food should be cheap and easy, and that eating distracts from and limits the time needed to "enjoy" life robs us of that very enjoyment by degrading nutrition and discrediting the benefit of having an equitable relationship with whole foods. A dismissive mentality spawned by advertisers to draw our attention away from value-based decisions about what food we eat ultimately destroys the economy and limits what is available for us to eat.

Current business practices in this entire network of interactions are guided by our cultural values, and those cultural values are reinforced with each action every individual makes in their daily lives. Once again, we see the interdisciplinary connection between valuing food, the food economy, and politics. Attempting to make changes from the top down is difficult without strong leadership and a refocus on values, a fact that experts recognize.[112] Change might seem like an unsurmountable challenge due to our culture's commitment to personal freedom, yet freedom is not what prevents us from changing. It is what gives us the chance to trigger change today in ways that have a real impact instantly for those who want it.[71] Educating ourselves about the implications associated with a choice makes it easier to make a different choice and break the cycle.[10,50] When choices are made based on the values that are most important to us, rather than on the illusion of convenience, the overwhelming evidence has shown that dedicated individuals can inspire their communities to work together for the benefit of all.[6,21,31,32,37,55,56,57,91,92,93,101,110]

Understanding the chain reactions like the feedback systems demonstrated in the chart earlier in the chapter is important and can help us predict how our local community can be most easily influenced.[96] We live in a free society where the individual has a voice, and that voice is heard through the choices one makes. Personal freedom is something each individual must claim every day. It demands the individual be knowledgeable and take ownership in support of their values and health. It requires the ability to visualize how choices made today impact ourselves and our communities' tomorrow. As social scientists, Magnus *et al.* eloquently phrased it "It is important to acknowledge that food is not only a material condition that exists in people's lives but is a vital element in their life-worlds."[96] The stronger our dedication to change, the more change we will see spreading out from our center. When we join forces with others who share the same values, our collective power grows.

The real measure of a life lived well is time spent enjoying family and friends, and participating in activities that inspire passion, creativity, and accomplishment. The responsibility to achieve this level of peace in one's life is assigned to the individual and any failure to succeed is not the fault of the community or government, but political leadership should at least work to support and encourage that goal instead of erect barriers in attaining it. What I have found in my research is that promoting public health and promoting a strong economy can be complimentary missions that reinforce and protect each other. The consumer must combat their apathy and resist the idea that one must choose between a strong local economy and health, that greed is more important than health, community, and the environment, and that enjoyment of life is solely dependent on wealth. There is no question that our culture forces the individual to compete each day for financial survival, but the level of engagement in that competition is up to us.

It is easy to perceive the system as too big and too established to be changed and modified, but it is a myth. The market is fickle. Almost overnight, products explode in availability when the consumer shows demand, and others fade just as quickly when a trend ends. This is an amazing asset. Consumers have power. I have read plenty of sources whose mostly well-intentioned opinions are obviously born of a perspective that has never had to choose between a meal or paying a bill; I am not one of them. I have wondered where my next meal would come from, and although I have achieved a modest level of comfort over time, I know how precious and fragile that comfort really is. I know that what I suggest consumers do to trigger a change is not an option for all initially; instead, those who can must pave the way for change so that others may be lifted.

Most importantly, we need to realize that change does not mean that the economy must suffer, rather with change the food system economy will grow. Encouraging smaller local farms will keep our food dollars circulating in our local economies and provide job opportunities. Food processing companies do not need to be eliminated; they simply need to be refocused to abide by certain ethical standards that the consumer demands. They may choose to make these changes voluntarily or may need encouragement via regulations. Consumer facing industries like restaurants, groceries, and convenience vendors are already primed to adapt. There are countless unique business owners who may benefit from re-examining their products and providing the consumer a clearly healthier option, and they will be rewarded with consumer loyalty as a result. Arguably, the recycling and waste management industries have the potential for vast growth in several different parts of the food system chain. Using technology to increase efficiency and reduce overall waste would bring value to our nation and our global community while ensuring each individual has access to nature in whatever form best suites them.

HOW TO MAKE HEALTHIER FOOD CHOICES

I have talked a lot about making choices in this chapter, so how does one begin? Because each person's diet is unique, every path will be different, but observing another's experience can provide courage and confidence. For me, I decided to make a significant change while researching this book. Here is how I made the switch to traceable meat and why. I had never been the kind of person who had to have meat at every meal. I ate meat or fish roughly four to five times a week. While researching this book I read about the hormone cortisol, which is a stress hormone that contributes to heart disease.[113] I learned that cortisol present in meat is absorbed directly into our bodies when we eat it, and the levels of cortisol found in commercially raised and harvested livestock are dramatically higher than the levels found in meat harvested respectfully.[21,37] I also learned about the significantly negative impact raising commercial industrial livestock has on the environment.[23,58,109] Next, I began to check out the prices of locally raised and respectfully harvested meat. Locally raised meat is often referred to as "traceable" because you can interact with the farmer at a personal level. Farmers raising small herds want to share their passion for raising livestock, so they tend to be transparent, even eager to explain the quality of their product as a result of the specialized care they provide. I was able to find a farmer who is located less than ten miles from my home who raises a very small number of cattle each year on a completely natural grass-fed diet. I was able to purchase one-fourth of a cow for a fraction of the cost of purchasing ground beef and steaks at the grocery store! I did have to invest in an extra freezer to hold the meat, but even with that cost, I was still well ahead of what I was spending on meat for only a few meals a week at the store. A family that consumes more meat would see even greater savings.

Locally sourced eggs were easy to find where I live, and their cost is comparable to the grocery store so that was an easy choice. I am also fortunate enough to have a specialty grocery store in my town that only sells locally grown and healthy foods, and they carry chicken and pork from farms within one hundred miles of my home. I researched those farms and decided their meat fit into my "traceable" plan as well. I also found a farm where I could purchase pork and lamb in bulk like the cow, but I simply don't eat enough to buy that much. While the traceable chicken and pork at my local store is more expensive than buying bulk like I do with the cow, I still had not matched the budget I was spending prior to this change, and with the freezer purchase I was able to wait until those other meats were on sale and buy more at the discounted prices. I am fully aware that not everyone has the ability to essentially pay their annual meat budget up front tomorrow. For those who have the resources, the first step is understanding why this is important: to improve health by reducing the amount of cortisol ingested, to support local farmers in the community, to save money in one's food budget, and as an added bonus—increasing the enjoyment of meals with meat that taste better. Planning about how to afford the change develops once a choice becomes important to the individual.

By now, my partner and I have been eating traceable meat for over three years. I enjoy meals more than I did before, knowing that I have reduced my carbon footprint and supported my local economy. In addition, both the taste and physical benefits of healthier meat are all positive re-enforcements. I have also talked to my local restaurants about where they get their meat from so I know when to order a vegetarian or seafood dish, or when I can enjoy traceable meat when I am out. The pessimist might say my actions have no impact, but I disagree based on a multitude of sources I have already referenced and my own experience. I talk to my friends about my choices and help pass on contacts for farms when asked. My close friends may often joke about my "knowing the name of my dinner," but it means that the topic is being discussed and the choices of other's may sometimes be influenced by mine. Those choices then grow exponentially. Importantly, in conversations, I always counter objections about the cost of time and effort in consuming traceable meat with the expense taken on my health of giving in to convenience. Keeping the values that are important in focus makes it easy to maintain my new habit and helps others evaluate what is really important to them. Although my routine is different, it isn't more time consuming now that I have built my network, and my personal food system chart places me a lot closer to the farmer.

CHAPTER 6

FOOD, RELIGION, AND SPIRITUALITY

I have three strong food memories from when I was a kid—two positive and one negative. The negative memory has resulted in my refusal to ever eat rabbit or lima beans since, despite how delicious rabbit may be. The other two memories have impacted me in much more subtle ways over time and have been tugging at my mind as I have researched and written this chapter. The first childhood memory is of neighbors who lived down the street, a pastor in a local church and his wife. His mother (or his wife's mother) was a Polish woman who would visit annually, and when she was in town, we were invited over to help make perogies. The "grandmother" would tell stories about where she grew up in Poland, and how she learned to make traditional foods from her mother and grandmother. For her, food was directly tied to ritual, family, celebrating events, and sharing with others. I grew up poor; there is no doubt we were invited to help make the perogies and take some home as a form of charity by the woman and her family.

Ever since that experience I have never liked to eat frozen perogies from the store; they have no flavor and don't cook up with the right texture. It's kind of ironic to look back and think about how a gesture of charity turned me into a snob who won't settle for an inferior imitation food product. For decades, I didn't even think about perogies; then when I went back to school as an adult learner and started thinking about food in a different way I remembered back to that time, made a batch of perogies, and froze them to see if the taste memories were real or romantically idealized. They were a hit with all my friends. Now I make a big batch every six to eight months, and they are a fast easy meal or side whenever we want them. The best part is I know exactly what is in them. You may be wondering what this has to do with the chapter topic, but please indulge me, and I think the link will become clear.

My other memory is of my best friend as a child, Naheed, whose Muslim family was from Pakistan. I honestly had no idea what that meant when I was eight or nine years old, growing up in a rural Western New York town. I don't think my mother knew that much about the Islamic religion either, and there was no Internet or electronic devices to look things up with (can you imagine?). One summer, I remember Naheed's family celebrating Ramadan. I was intrigued by the religious custom, and I wanted to fast like my friend. My mother was very concerned that fasting wasn't healthy for a child, and that I would feel weak or be grumpy. She said I could do it for three days only, and if I acted up, I would be forced to eat. Naheed gave me some of her traditional clothing she had grown out of, and I felt like I was adopted into her family. For three days, I fasted like my friend Naheed, our play outside was subdued, and after sundown, I was invited to join her family for their nighttime meal. I am ashamed that I didn't learn much about her religion from the experience, but I was exposed to foods I had never heard of, and they were delicious! Perhaps they tasted so good because I was so hungry, but I think in general the flavors were simply exceptional and like nothing I had ever tasted before. Researching this chapter topic has caused me to really consider that experience in a way I didn't have the capacity to when I was a child. I am grateful to Naheed and her family for that memory that has stuck with me many decades later.

These two experiences from my childhood share a connection to both food rituals and religious practices even though the religious aspect wasn't a direct influencer for my own personal path. Instead, the food experiences, and engaging with others during preparation and enjoyment of a meal, hold important connotation and nuance. The social connection and intimate relationship we share with others surrounding the act of eating is an important part of our lives as humans.[11] In a wonderful book called *The Magick of Food*, Hughes introduces the author and the text eloquently when he says, "Food and drink do not simply sustain and nourish the body; they have long been associated as sustainers of the spirit, family, and community."[20] Food does more than just provide the fuel for our physical bodies. This chapter will examine how foodways guided by beliefs enrich our engagement in the present moment and help define our place beside other living beings in the context of the larger living planet. The idea that religious beliefs or rituals involving food influence nearly everyone's daily lives, many times unconsciously, has become exceedingly apparent to me while researching foodways from this perspective.

To start, I need to state explicitly that religion and spirituality are not the same, nor are they interchangeable. Religion is defined as "set of beliefs concerning the cause, nature, and purpose of the universe, especially when considered as the creation of a superhuman agency…usually involving devotional and ritual observances, and often containing a moral code governing the conduct of human affairs…. generally agreed upon by a number of persons."[114] In other words, religion is a set of concepts or doctrines often referred to as theology that explain or provide purpose for existence, and that a group agrees is true. Spirituality is defined as being "of or relating to the spirit or soul, as distinguished from the physical nature."[115] Spirituality simplified is how an individual perceives themselves in relation to the living energy all around them. An individual may be influenced by religion or spirituality in their lives, or both combined in varying proportions.

The commonality between religion and spirituality is that both relate to and help shape our mental and emotional states. While our physical existence is limited by various factors including social, financial, and health status, our emotional and mental capacity has the potential to set us free.[116] We can dream. We can experience joy as deeply as we choose. Our ability to feel connected to others (human and nonhuman) helps us feel less lonely and afraid and provides us with a sense of well-being regardless of our physical boundaries. A devout practitioner of various theologies may choose to let go of physical entanglements, sometimes referred to as a vow of poverty, in an effort to achieve an enlightened state where they feel complete and fulfilled in their existence as a living being. The same result can be experienced by a strongly spiritual person as well; they simply have different language for the actions they take on their path to enlightenment. Food can act as a tool, a gift, a daily meditation, and a form of worship as one strives to find meaning or connect to something bigger.

Food is always a part of one's journey as our existence literally depends on the energy we receive when we ingest it.[2] For that reason, most religions have dealt with food directly in the establishment of their rituals. People who don't claim an attachment to an organized religion but recognize the importance of food have developed secular rituals such as vegetarianism and the Locavore movement that have underlying themes related to spiritual concepts.[21,103] While many scholars and experts tend to deal with food security and health issues by focusing on tangible solutions like policies and nutritional education,[6] I have found through my research that beliefs systems and personal convictions based on moral and ethical principles play an equally significant role in what we actually do.[103] A good example can be seen in how religious groups are often the coordinators for food aid programs. Charity or community programs that increase access to food for those in need have a direct impact and provide an emotional reward for those facilitating the giving as well as those being supported,[116] yet this band aid does not address the institutional causes that instigate food insecurity.[6] The solution to food insecurity must combine government actions with community outreach programs to be effective at solving the problem in the long run.[8,46]

As a society, we acknowledge there is an issue with food security in some communities, we complain that food is too expensive even when we have the resources to buy it, and we practice food rituals with varying degrees of awareness. We do all these things while complying with an economic model that inherently degrades food quality and our ability to continue to access fresh food in the future.[6,8] We maintain a dual consciousness between our beliefs and our actions; this is why an interdisciplinary perspective is crucial to really comprehending our relationship to our food. To truly understand our foodways, one must better understand the mental and emotional components that both drive our habits and conflict with our values.

In the culture of the United States, I think this is difficult to holistically self-evaluate because we tend to dismiss emotions as a weakness. We routinely promote profit over feelings of peaceful wellbeing, or we attempt to buy serenity versus looking inside ourselves to identify why we are unhappy.[8,21,50] We ignore food as a gateway to experience joy in the fullest and see it only has a nuisance that can't be avoided and a commodity to profit from.

Being spiritual represents a resistance movement to our cultural norm. Spirituality embraces the human emotional component and strives to align the ethereal self with the physical self to eliminate conflict and find harmony. Practicing spirituality requires consistent attention, so paring it with food practices makes sense on multiple levels. Because most humans need to eat regularly multiple times a day, food can become an awareness trigger for self-contemplation, a testament to one's beliefs, and a way to connect to the energy shared by all the living beings on our living planet. Choosing foods for their freshness and nutritional value demonstrates consideration for one's own health and well-being, the quality of the food item, and those who raised that food for consumption.[50] A spiritual approach helps highlight that food choices, positive, negative, conscious, and unconscious impact both the individual and their community at the same time.

Inherently, a spiritual mentality tends to extend outwards toward one's community rather than inwards as a selfish mindset. While it is a common argument that the individual is at a disadvantage when they make community a priority over self, studies show that community strength benefits all individuals involved.[6] This uplift through a tight knit community can also be seen in religious groups.[102,116,117,118] Being a part of a congregation provides a feeling of security through emotional and physical support. Many congregations actively work to provide assistance to their members and within the larger communities where they live.

In my area, Catholic Charities provides resources to soup kitchens, school lunch programs, and many other support programs with food as a primary asset to uplift those in need. The local college, Skidmore, also acts as an example of what is possible when students, faculty, and staff of various faiths come together to support community programs with food and other basic essentials during the winter holiday season as part of their Skidmore Cares event. Experts like Sack explain how outreach to the community through food programs is an important tenant in various Protestant religions and congregations all over the country.[118]

My friend Judith who served as president of Congregation Shaara Tfille for several years has explained that specific tenants within the Jewish faith like tikkun olam—repair of the world; mitzvah—a good deed done from religious duty; and tzedakah—a moral obligation to charitable giving are important mandates from God.[119] She explains "These themes guide us to participate in a variety of outreach activities that provide food for those with food insecurities."[119] The Sisterhood within her congregation donates to various programs such as "SnackPack" that gives 170 children in our area bags of groceries to help support them on weekends when they don't have access to school lunches, and Code Blue, a local homeless shelter. For many religious groups, outreach is not limited exclusively to members of the congregation; the entire community is supported through their actions as part of their belief systems. The giving is not necessarily tied to expansion of membership, but instead, an action that bears witness to one's dedication to the doctrine of their faith.

The giving of food to those less fortunate is not the only way some religions look at food through theological lens. Many religions have distinct rules about foods to avoid or combine. Judith cited specific biblical passages, Leviticus 11 and Deuteronomy 17, as rules from God that define kashrut practices. She outlined three ways keeping kosher supports her faith: it helps keep religion "alive," present, and active in one's daily life practices, it helps define what is healthy and what is not, for example, fast food is not good for the body or soul, and it helps strengthen one's relationship with God through demonstrating devotion. Robinson, Professor of Religion and the Environment, writes about her study of food culture and echoes these same ideas when describing the halal rules for food in the Muslim faith.[103] She describes how a business in Chicago has made it their mission to aid people of faith in reviving halal food traditions that may have been lost in the process of assimilating into the United States' cultural foodways; another important aspect that will be considered later in this chapter.

Overall, it is not uncommon for organized religions to claim certain food restrictions or additions as dictated by their God to help maintain the health of the faithful. Cindy, a labor and delivery nurse who has studied and practiced several religious theologies on her personal path of faith, noted that ancient religious texts contain useful and accurate instructions for health of the earthly body.[120] In the Quran, pregnant women are instructed to eat dates. Cindy has witnessed how consuming dates has shortened the period of labor and eased the related pain for her patients.

There are countless ways that religious texts influence the diets of the devoted, and sometimes, overlapping reasons for the rules including but not limited to demonstrating obedience and improving health. The Catholic practice of abstaining from terrestrial meat sources and substituting fish on Fridays is just one of many more examples that the average American might be familiar with, yet the connection between food and religious devotion extends well beyond for some groups. The concept of eating according to rules dictated directly by God has been expanded on and adapted by certain Christian sects. There are multiple examples of congregations whose leaders claim that a vegetarian diet, vegan diet, or other restrictive food practice like a "raw foods" diet is supported by the Bible, and that the purpose for this food rule is to attain a physical purity above those who do not believe.[103]

Here, the focus shifts inward not only using dietary abstinence to demonstrate faithfulness and devotion as well as protect the earthly body that has been provided by God, but to make oneself cleansed and ready for the reward of afterlife. While the reasons for these food rules identified by sect leaders are an implicit translation of guidance from religious texts and therefore are different than kashrut or halal rules dictated explicitly, the overlying general intent for both is to honor the physical body as a gift to be respected and to provide a daily reminder and reinforcement of faith. Food is also integrated into worship ceremonies for many Christian and Catholic denominations in the practice of communion; food is used symbolically as a physical embodiment for connecting to their God. Ingesting the anointed food confers acceptance of cleansing and reaffirmation of faith; yet another form of claiming a sense of purity over others.

For modern worship groups that observe specific food rules, the evidence to support their special diets is sometimes derived from a combination of biblical passages and scientific evidence that mutually support each other.[103] Science is a mainstay in our culture and using science to support and articulate how an individual should eat in order to honor their beliefs helps reinforce the habitual practice. This strong interdisciplinary connection between food, science, and religion surprised me at first, but the longer I thought about it, the more I realized that for some in the United States science has replaced religion because it provides documentation and confirmation rather than requiring blind faith. Many others have found a way to blend their belief systems with scientific understanding regarding food, medicine, and more; to me, it appears a necessary mechanism for living in a technological and evidence-based world while remaining connected to the intangible parts of ourselves that we can feel, but that science can't define.

Jewish, Muslim, and Christian faiths all share a common origin resulting in some general similarities with regards to having food rules that focus in myriad ways on honoring the bond between them and their God and observing rituals as part of the path to a future that is the prize for being faithful in the here and now.[103] Other belief systems perceive their place in the world differently. Although very distinctive in specific doctrines, both Hindus and Buddhists generally tend to have a more holistic and cyclical view of one's current lifetime, and food practices of followers reflect that. With a primary focus on being the best version of oneself in the present, the individual is encouraged to be more respectful and aware of other living beings sharing the space we exist in now.[52]

Thich Nhat Hanh, respected Buddhist leader, outlines several precepts for study and practice.[121] Many of these precepts involve describing how one should interact with the world around them including one's relationship to food. He notes that it is important to seek to recognize suffering rather than ignore it, "do not accumulate wealth while millions are hungry," and "do not mistreat your body."[121] These are not explicit rules for what to eat; instead, they are guides by which one can reasonably construct their foodways well beyond avoiding particular foods. The practitioner is not only challenged to eat in a way that respects the food they are eating by considering that it was once alive (even vegetarians eat formerly living plant beings), but also where it comes from, the impact of the agricultural practices used to produce it, and if the farmer or grower was paid a livable wage. The eater is further charged with guarding against personal gluttony and sharing with those less fortunate. In this way, every meal becomes a lesson in awareness and an opportunity for growth.

The Buddhist practice is in some ways like Native American food-ways with regards to a larger active perspective in the moment. Native American practices revolve around respect, awareness, and gratitude for the nutrition provided by nature.[28,29] For tribes in the Northwest, staple food items such as salmon are the godlike providers that created their people, and also represent the embodiment of their ancestral leaders.[103] Salmon as a food is a gift; that gift continues to be provided only as long as the rituals associated with it are honored, and the environment the salmon needs to survive is protected.

While it is quite literally true that the salmon could die off if the environment is damaged, the idea of protecting the home of the salmon has a much deeper meaning for Native Americans. Native Americans demonstrate a high level of spirituality in their belief systems, recognizing that the spirit or soul is infinitely connected to the natural world, and that the spirit or soul is not exclusive to humans. All living beings contain energy that transcends the physical; even rocks carry a consciousness.[104] As a result, food is a critically important interaction point where the spirit world blends with the physical.[29] This mentality fosters practices that cyclically protect the foods the individual needs to live well and healthy, and the environment that continues to provide the food.[28]

Holistic perspectives of foodways by some groups truly embody the idea of worship. Care of self, care of community, and care of the environment are not actions naturally opposed to one another, yet for some reason aren't always the standard in our everyday lives.[8] The melting pot culture of the United States has in some ways degraded the various and diverse belief systems that people have brought with them from all over the world as they have attempted to merge into one cohesive society.[9,103] While the culture of our country is a blended concoction of traditions today, the ideology held by the European settlers who established our government formed the base set of principles that became the policies governing agriculture and the economy.[8]

Here is another interdisciplinary connection between food, religion, politics, and economics. We elect government officials in part based on their integrity which voters measure through their show of faith in a recognized theology. Presumably, voters do this expecting that religious devotion implies a level of ethical or moral fortitude. Yet a key tenant in governance is separation of church and state. Freedom to worship as one chooses is an ideal we demand. So is choice of spending. Capitalism as an economic model promotes innovation and manipulation of food products at varying levels of intensity in order to bolster the volume of and dollars spent on purchases overall.[8,15]

Cash flow through the economy on food purchases can be grouped into two general categories. Purchases can support local small businesses providing a food product that requires skill to prepare, is supplemented with a memorable experience, or simply saves the eater time. Purchases can also be funneled to a limited few who have used economy of size and inexpensive, storage friendly, man-made artificial ingredients to produce profit rich imitation food products. Convincing consumers to desire new food creations and thereby possibly reject traditional food rituals was achieved through the development of an entirely separate component of the economy—the advertising industry.[6,8,15]

Conversely, not all consumers are encouraged to forget food rites that have a conscious link to their faith; advertisers can help identify which foods fit religious dictates, too. Although this has led to some questions about the integrity of the certification process of halal or kosher foods,[103] consumers are able to easily identify which foods fit into their religious precepts. Consumerism has also driven an expansion of vegetarian, vegan, and organic food options through demand.[15] One can argue that such diverse choices allow consumers to eat according to whatever food rules are comfortable for them. The reality is that fresh whole foods which are the ingredients in traditional dishes are ignored by the marketing industry, likely because eating them involves time and preparation, which is hard to sell in a culture conditioned to renounce everyday meals as valuable events that hold meaning.

Examining various media sources could lead one to assume we rarely think about our foodways at all unless we are striving to lose weight or deal with a health issue.[8,11,50] We have the ability to enjoy connecting with others at a special mealtime, but we don't routinely think about it as a specific health maintaining action, or what might be missing when this isn't a regular practice. I think even fewer people perceive everyday meals as an opportunity for an extraordinary spiritual event to connect with nature.[13] From my experience, most people enjoy knowing a meal has been made from fresh ingredients and with care, but the subject of intent on behalf of the food source is discomforting for many. Those who embody strong spirituality tend to find the theory of a connection through intentional energy transfer a surprising but reasonable concept. Alternatively, the idea that our food sources possess the capacity for intention gets mixed responses from those who identify as religious or claim no association with an ethereal creed.

This is perhaps the most difficult part of examining food from the perspective of religious and spiritual faith. Faith is "belief that is not based on proof."[122] Faith and beliefs are inherently emotional and, as a result, tend to trigger a defensive response when challenged, especially if the individual lacks confidence.[6] Faith is valuable and necessary in our daily lives but, like everything else, must be in balance to aid us in interacting with the world around us. If simply asking someone to consider the intent of the living beings providing our nutrition causes a negative emotional response, how can we as a culture promote the exploration of the world around us related to our health that is still beyond our current comprehension? This question helps to frame morals and ethics within the larger topic of religious beliefs.

Depending on what interpretation of religious texts one subscribes to, they may perceive themselves as above or destined to control nature, while others may feel a responsibility to protect and honor nature. An ethical approach to food implies that consideration and thanks are due to our food sources.[13,21,28,29] Morals, defined as "the principles or rules of right conduct or the distinction between right and wrong"[123] should effortlessly overlay the actions of a religious or spiritual person striving to embody their belief system. Instead, viewing other living beings as having equal rights to human's remains outside mainstream acceptance.[124]

In my life, I have observed countless examples of those who flip back and forth between living according to a theology in certain spaces and completely opposite it in others, and those who justify their actions by what seems like a perversion of religious tenants. This is not a statement of judgment; it is simply an observation that our beliefs guide us, but temptation to follow cultural norms is also a factor. Our culture of consumerism thrives in part because of our attempts to overcome the guilt of acting in a manner that does not align with our beliefs.[11,50] Specifically related to food, we are willing to pay for all manner of quick fixes versus changing our diet and eating habits to honor the magnificence of our bodies.[8,11] At holiday dinners, many pause to give thanks to their God for the meal and the companionship, but not to the food items themselves. I don't think it is sacrilegious to also thank the actual food for its sacrifice. Thanking another living being is not the same as worshiping it; humans thank one another every day without violating religious beliefs. The act is simply courteous.

Perhaps science is the culprit for our perception of living beings that provide food as undeserving of gratitude or lacking the intent of generosity. It is generally accepted that only humans and possibly a few high functioning animals are sentient. Darwin's theory of evolution categorizing humans above other life forms has been widely accepted instead of alternative suggestions that all life forms operate in relation to and in concert with one another with diverse but vital obligations to the success of the whole.[48,85,86] But even Darwin questioned the possibility of sentience in nonhuman living beings, and scientists, philosophers, and literary critics are starting to seriously explore how our current accepted norms might be completely mistaken.[12]

While various theologies deal with the evolution/creation conflict in myriad ways, few leap to embrace humans on equal footing with food providing beings. Perhaps this loose consensus forms the basis that limits the number of scientists willing to pioneer forward and consider if the measure we use to award sentient status to nonhumans is accurate or needs to be redefined.[12,80,85,125,126] Studies have shown plants can communicate, remember, and learn; recognizing that should increase the belief of religious people in the power of their Gods, not challenge their faith. From the spiritual perspective, a greater pool of intentioned energy existing in the world increases the strength of all who work together sharing that energy.[127]

Somewhere between using science to explore what our food provides to us in addition to complex chemical compounds for nutrition, and belief in something that gives meaning beyond our mechanically physical existence lies what takes place each time we eat. We don't have to completely understand how nutrition works; for centuries, ancient people survived and thrived without specific knowledge of carbohydrates, dietary fiber, antioxidants, and linoleic acid. The difference is that our ancestors had stronger food rituals linked to belief systems.[20,103] The act of eating was often shared in a reverent practice. Today's society has traded a holistic approach to diet for a reductionist and unbalanced focus on nutritional components[2,21,50] and eliminated the honoring of mealtimes as sacred in everyday life.[52]

We may still gather for special occasions to honor a holiday or a milestone, but daily meals are often consumed on the run and without any fanfare. Fewer still practice regular rituals like "kitchen witch" Raven describes that look to celebrate good health, give to those in need of healing, and commemorate those who have completed their lifetime journey as an ongoing routine.[20] Raven says intentional gatherings focused on symbolically acknowledging the blood energy that flows through us create awareness, "Eating reminds everyone at the table that we are not dead yet", "healing meals reduce stress in sick and dying people" allowing for one to resolve conflicts and say goodbye, and "dining together can be a time to foster apologies and offer forgiveness."[20] It is hard to argue the benefit provided by shared meals between friends and family that connect us to each other, yet it isn't a priority in our culture. Perhaps that is why giving credit to the contribution of the food itself for the benefit of the participants is so beyond our comprehension.

What sticks out to me in the years I have spent researching food is that the meals that hold the greatest significance for us with regards to connecting with others are those prepared with care. Whether a host provides all of the food items for their guests or guests are each charged with contributing, the food at special gatherings is given thought, attention, and time. What is served does impact the moment. Dishes made from whole food ingredients and favorite recipes convey something intangible. Even at a casual picnic, with cabbage salad, potato salad, deviled eggs, and kabobs, these dishes transmit a sense of generosity that would not be replicated if everyone brought a premade processed food item from a box. The food itself does contribute to the experience.

Do we resist the idea of sentience in food sources out of guilt that a life has ended so we may eat? For me personally, envisioning that the food I consume is giving spiritual energy to me intentionally in addition to caloric is comforting. I am a gardener and I have spent decades nurturing perennials and being rewarded by their beautiful flowers. When I started on this food exploration journey, I began mixing in vegetables to create an edible landscape in my yard. I don't have a traditional vegetable garden with neat rows. Instead, the peonies, daylilies, wisteria, daisies, and others draw me into the garden and encourage me to keep an eye on the snow peas, green beans, garlic, tomatoes, and eggplants so I am always alert to what is ready to harvest. I try to mimic the Native American tradition of giving an offering to the plant that is providing me the tasty meal.[29] While I am eating, I strive to be consciously grateful for all that is becoming a part of me, physically and energetically, and I feel different, healthier, now than I did before I started this practice.

Even if my experiment of visualizing an intentional energy transfer from my food has no documentable physical benefit, it does influence how I feel, my perspective regarding the world around me, and my actions associated with sharing the gratitude and value provided to me by food. Overall, my emotional and mental health is improved. My experience is not unusual; emotional and physical health and well-being as a result of connecting with nature is well documented.[16,17,78,79,86] This is an important point. An exercise that improves the self through an act that is necessary and frequent like eating, is available to everyone and benefits the entire community as it grows.

I realize that fresh whole foods are not always easy to access for underserved areas through conventional outlets like grocery stores, but there are many examples of urban gardening groups that have found ways to provide at least some options, and as demand grows more fresh food choices will develop.[33,34,35,100,128] In any case, the act of thinking about food differently can be done using whatever fresh food is available now and expanded as new sources are identified. The first step is changing one's mindset with regards to food to heighten one's well-being and enjoyment of life.

Ultimately that is what spirituality and religion provide, an enhancement, something that is individually experienced but shared by a group and gives meaning to our existence. Beliefs inspire positive feelings of joy, purpose, value, and assurance to balance the negative emotions often caused by our societal routines. Food plays a part in this process every day whether we recognize its importance or not. Religions were developed to include food rules and rituals for the explicit intent of keeping beliefs a part of our active daily lives.

Our choices regarding what we eat and how we eat it demonstrate how we measure our self-value regardless of whether we acknowledge it and act with purpose, or coast subconsciously through daily meals.[11] The effort we invest in preparing, sharing, and nurturing ourselves with food is a critical piece in each individual's foodway. Some families maintain rituals and rites that have been passed down through generations, while others have adopted practices that fit newly formed beliefs. Still, we all eat, and we all have habits with regards to what and how we eat, the roots of which developed centuries before we were born and are carried forward with us daily.

CHAPTER 7

FOOD, CULTURE, AND ETHICS

Food considered from the perspective of culture and ethics gives me the most hope compared to the other disciplines already discussed, and it is intimately intertwined with each of them. We have looked at how our values are the foundation on which our culture is built. Despite our actions that may suggest corruption in our values, I think once we stop and consider what we stand for, the majority of our country's population does possess a strong, positive, core value set. How food impacts our health has been established for decades. Desire for quick fixes and falling for marketing that seductively pushes the unhealthy are symptoms of our culture; we have been told we deserve rewards without any personal investment, so we ignore the knowledge validated by experts and science regarding what our bodies need to perform at their best.

Scientific exploration is most often driven by money and power. Few scientists have the luxury to explore based on curiosity; James Lovelock, father of the Gaia Hypothesis, was a rare exception and even he admits he had to make concessions at times.[87] I wish more than anything that innovative scientists who aren't afraid to explore and challenge accepted ideas about plant consciousness weren't a rarity,[80] and that many more would step up and consider the relationship of energy transfer beyond the caloric between our food sources and us humans. But the industrial agriculture monolith has spent decades trying to sever even the idea of that connection, and who can independently afford to fund such brave work. Biologist Gagliano, who has published some of the most exciting new ideas about plant sentience in the past decade, demonstrates how much we still have to learn and that we must let go of some assumptions in order to get there.[80] The lack of scientists focused on this topic reflects the limited funding available. Certainly, most individuals have little influence in the direction scientific exploration takes; instead, we mistakenly trust that there is a baseline of integrity in the cultural trends established by the powerful.

We have only slightly more voice when it comes to political and economic policies. We hire (elect) leadership to govern us, but the information we use to choose those leaders and their effectiveness once in said position, is more influenced by money than anything else, as is the food section of the economy where money is solidly controlled by the industrial agriculture machine.[8,23] I have tried and have no idea how to wake my representatives up to the fact that I expect them to be effective and find paths forward to improve food security, sovereignty, and minimize climate damage, so we all have food in the future. Instead, it appears they spend much of their time acting like children on the playground obsessed with a marathon game of dodge ball. As a result, policies that should work to protect farmers, farm workers, and the growing of fresh whole foods that would benefit the entire population, always seem to trend the opposite direction toward making the giants in conventional industrial agriculture even richer. Apparently, our government leaders mistakenly trust in the integrity of cultural trends too.

Our participation in religious or spiritual ideologies can support a positive food culture by ritualizing practices and expanding one's concept of their place in this moment. Ethereal beliefs can reinforce the wisdom of how diet impacts health and inspire communities to provide security nets for those who don't have enough. They can also be manipulated to exclude based on measurements of purity set by human leaders who are fallible. Only a few belief systems remain today that honor nature for the daily miracles provided that nurture nutrition as a primary focus.[20,28,29] Unlike other disciplines, we maintain full agency through the freedom to choose what to believe and how much time to invest in practicing that belief, including whether the ideology recognizes food as miraculous. It is this discipline that provides substantial historical evidence regarding the potential we have to influence our culture, and our puzzling tendencies to surrender that potential without protest.

All of the disciplines discussed here inform and are informed by our society's culture. This is where our power lies. The response of the economic sector to burst forth with products to satisfy every popular demand triggered by pseudo-science or miss-interpreted micronutrient discovery is just one example. The prevalence of safety net programs for the hungry that are funded entirely through donations is another. Individuals, joined together and focused on a topic or cause, drive our culture. I feel how easy it is to look at the disciplines involved in our food systems and feel small and insignificant. At the same time, I have seen how my passion to inspire others through food gives them energy to be great in their own way. We may not be able to change our food systems from the top down, but we can shake the foundation it is built on and make demands on how it is rebuilt.

Who we are and how we see ourselves has many layers depending on the context of the moment. Personally, I see myself as a dreamer, an artist, and an investigator. In my home, I am a partner, an encourager, and a cook. At work, I am a peer, a mentor, and a trusted resource. In my community, I am a consumer and a contributor. In our country, I am a voter and a taxpayer. Even I don't naturally identify as an eater, a receiver of nutrition and life force at any of these scales, even though that is what allows me to be all these other things. We forget, or never learned, to snuggle in the embrace of our nutrition providers and feel the intense strength being given to us when we eat; the energy that makes us capable of changing the world. As a result, when the lens widens most of us tend to perceive ourselves as shrinking in personal influence. Our definition of community is narrow, consisting primarily of the people we engage with and where our actions produce results we can see. All other living beings, human and non-human, exist, but not in a tangible way.

It is this tendency to isolate ourselves that makes food culture so difficult to conceptualize. We are a part of our culture, defined as "a particular form or stage of civilization."[129] through whatever practices define our lifestyles. Our culture is also part of us, as "the quality in a person or society that arises from a concern for what is regarded as excellent."[129] It is a constant feedback loop. It mirrors the energy transfer cycle between the geological environment, the living beings who become our food as well as the ones who support those food providers, humans as the consumers of food and caretakers of the environment, and the elements and verve contained in all living tissues that return to the environment at the end of each life cycle so that new life can form.

There is no end point or beginning to the cycle of life energy transfer, and the same is true of our culture. I am an individual whose actions, values, and ethics, define my culture while I influence those around me. My opinions are also shaped by and measured against what is considered normalized in our society, locally and nationally. All aspects of all actions fall within this cycle; both conformity and resistance to it. My personal culture determines my strength and ability to lead or follow; to instigate change or add momentum to an already flowing tide. We can resist participating in our food culture, or actively fight to change it, but even in those actions we are part of what shapes the overall picture.

For this reason, every individual owns some responsibility for the millions of people who suffer from food insecurity every day in this country. Every choice we make regarding what we eat impacts others. If we eat most of our meals at fast food locations, we are monetarily supporting their growth and prevalence. Conversely, if we use our food dollars to buy fresh whole foods at farmer's markets or CSAs, we support the expansion of foods available in these types of outlets. As more people shift toward a healthier eating regimen, the economy surrounding food will rush to supply what is in demand. An example of this is how the industrial agriculture complex has expanded into the organic food market, even if their effort has perverted the intent to grow "cleaner" food to some degree.[23]

While occasional charity can soothe guilt for having enough when another does not, understanding that every choice one makes when ingesting a meal or snack is an expression of culture creates a level of ownership for every individual. This awareness expands the opportunity to consider beyond food insecurity to food sovereignty. Conscious thought about what we eat ourselves helps us to identify with every other human being's desire to also have the right to choose their food based on their values. Either we recognize this choice as a basic right that everyone deserves, or we acknowledge that our own privilege to choose in this moment is provisional and may be taken away without warning. The food choices we make today, tomorrow, and this week, build into a pattern that has far reaching repercussions for both ourselves and our neighbors in the future.

We have only minimal control of many of the outside influences targeting our food culture. We hope our elected officials make sound decisions based on comprehensive information rather than surrender to the demands of the powerful and rich. We practice the food related rites of our faiths because our core beliefs are maintained in them, despite the fact that those rituals were created thousands of years ago in cultures very different from the present. The science of how nutrition impacts our bodies has not changed since we evolved, even though we learn new details about it regularly. While our command of the external is limited to reactionary, the frequency with which we eat allows us tremendous power to promote and polish our personal food culture and even the food culture of our communities.

As a result, food ethics and culture are intimately entwined; to adhere to an ethical standard, personal culture needs to prompt and shape community, regional, and national culture. Individual food choices set the cultural expectations or tolerances for "distributive justice."[7] Ethics expert Sandler states, "A system, policy or practice is …unjust if those who benefit from it do not also shoulder the associated burdens, or if the benefits of the system are unequally distributed without good reason."[7] Being actively engaged with our food choices forces us to consider, contemplate, and cultivate our values and personal culture every time we taste, chew, and swallow. It also prevents us from becoming complacent and allowing food insecurity to grow and more of our neighbors to go hungry, get sick, and suffer. Enough fresh whole foods for all doesn't have to mean less for each person, quite the opposite actually.

The fact that our country (government and individuals combined) spends billions of dollars on space travel, entertainment, and sports just to name a few, but allows millions of people to suffer food insecurity clearly demonstrates where our culture is heading even if it conflicts with our personal values. Our society has turned its back on subgroups within our nation, allowing them to go hungry and be malnourished,[4,6,7,130] and that action reflects as a stain on our culture as a whole and individually. Ironically, those who have been ignored remain and possibly become a more prominent feature that defines our society beyond stated ideals. Over time, our lack of regard for our food choices contributes to a feeling of personal non-consequentialism, and I think that more than anything else is the driving force that separates us from each other and our food sources.

You and I can be many things to one another, and it all depends on how we think about it. From a positive perspective, centered in confidence, we are fellow humans following parallel lives. Whether I know you and what makes you special, or if I have never met you before, we are the same even if we look different. We share basic needs to live a fulfilling life—food, shelter, companionship, and inspiration. In this mindset, it is easy to care about and want the best for you because your happiness is something I can relate to. I am curious about your preferences and reasoning behind them, and I seek to learn from you. I value you because I value myself; we understand each other's essential needs because we share them.

At the other end of the spectrum is the negative perspective rooted in self-doubt. Full of fear, I worry that I must fight to have enough. I am protective of those closest to me but see anyone outside my circle as my opponent, anticipating attacks even when there is no evidence supporting the suspicion. I am willing to hoard and defend my food even if it means some of it will go to waste, because if I share with you, then I might not have enough later. The only time I will give you the food I no longer want to eat is if I think you might be so desperate, and that enough of you have banded together, that I can no longer defend my pantry. I feel threatened, angry, and isolated.

Realistically most of us fall somewhere in the middle between these two perspectives. A wide collection of cultural influences seeks to normalize the adversarial relationship between you and me because it increases spending. Keeping up with the "Joneses," rating comparisons in the form of grades in school and salaries at work, sports match ups, and practically every movie theme ever, paint a picture that life is a constant competition. We might be aware that almost every social activity we engage in has some connotation of rivalry underlying the moment, but rarely do we realize how this negatively impacts our food culture. As we distance ourselves from one another and our food, we increase the negative feedback we feel without a clear understanding of the connections between the action and the reaction. To clarify the connections, one first has to take inventory of their own food culture.

Exploring how food culture is represented for the individual within their community is not a simple task. Culture is the most important part of our food system; it has the potential to drive the path of all the other food disciplines discussed so far. The problem is that food culture is intangible and diverse. It is easier to focus on health, science, economics, politics, even religion and spirituality, and how they all impact our food systems because they have a physical component or an established unified process. Meanwhile, food culture is a complex intermingling of personal, community, and national ideologies that require purposeful introspection to articulate. Defining one's food culture without consideration of the larger community provides an incomplete comprehension and appreciation for the repercussions of our actions. This relates back to the first chapter, and how the language and terms we use to define our food culture only complicate our ability to relate to one another and acknowledge our own worth.

As shown in the previous chapters, each individual perceives food related terms differently. "Good" food means many things to many people; it can signify nutritional content,[15] it might refer to taste, familiarity, or comfort,[9] it can represent storage and preparation attributes, it may be seen as an elitist qualifier,[4] or it might trigger feelings of contempt or persecution.[4] Defining what is "good" food is complicated by the unique set of experiences we have had from the moment we were born to present that form our perceptions. It is not surprising that beyond the difficulties describing food attributes consistently for the entire population, the confusion persists and intensifies in the struggle to converse about the fragility of our food systems.

Fragility can also be interpreted in multiple ways based on perspective. Ecological imbalance as a result of manmade processes and enhanced by climate change is only one aspect of the disease eating away at our food systems.[30] Food insecurity impacting millions of people, and disproportionately affecting minority communities, is another symptom of food systems that are dysfunctional.[4,5,6] Food sovereignty, the right of individuals to eat a healthy and culturally or religiously appropriate diet versus subsisting on a diet of highly processed food products that have been linked to health conditions such as obesity, heart disease, and diabetes, is yet another measure by which current food systems are failing.[24,25,26] To address the problem of a fragile food system, all parties must acknowledge and appreciate the variety of issues related to the food system.

Beyond that each person must recognize their unique viewpoint, which has been formulated based on the values they have developed throughout their lifetime for what it is; a limited perspective. Our interpretation of health and how food contributes to one's physical and mental well-being[2,22] and our exposure to science through education and advertising,[8,21] are just the beginning. Economics and politics are each complicated complex social systems that work in concert to shape both our personal and communal culture by controlling everything from how much currency we have access to earn and use, to how informed we can be about the manipulation of food products available to us, and how various food products are distributed to diverse communities.[7,8] Each individual also has a specific notion of celestial belief system that is part of their identity, and often those beliefs may dictate what should or shouldn't be ingested, as well as what rituals may be necessary as part of the process.[20,103] Because of these various factors, ethics expert Sandler points out that "(Food) preferences also can be inauthentic…(T)hey can be the product of external manipulation, rather than internal reflection."[7]

With a plethora of factors impacting us in varying ways and weighted by our individual cultural values, the differences between each of us seem unending. It is our overall culture that helps to unite us under one umbrella as a society. Every day, there is a constant exchange of personal ideology pushing outward and societal norms seeping inward with regards to food choices. Further, what direction our food culture grows toward is guided by our ethics. Ethics, "a system of moral principles, or the rules of conduct recognized in respect to a particular group or culture,"[131] are also embraced at varying degrees at the individual level. For instance, a declaration that food is a right for every human being and not to be used as a weapon by powerful people or governments, is driven by ethics and becomes a mission to achieve, if the majority of our communities consider it a worthy undertaking. Humanitarian organizations can promote the concept,[1,3] but without buy in, the message feeds no one.

Accomplishing goals such as this cannot be left for governments or businesses to initiate and tackle on their own; members within communities must insist that action be taken and provide guidance for what solutions should look like. Ethics expert Sandler points out that, "'Corporations' goal – and their fiduciary obligation in many cases – is not to promote public health, or even public welfare, but to maximize profits."[7] We who make up the societal body must set the cultural and ethical boundaries within which corporations must function. If we don't, then we have given permission for corporations to act in ways that likely won't reflect our personal cultures and will eventually infringe on the ability for all of us to eat according to our personal preferences or health needs.[13,27,28] By doing nothing, we allow them to put their profits ahead of our well-being.

The concept that community members are directly responsible to initiate change has been embraced by some. Within the current food culture of the United States, there are several themes that help individuals proclaim and promote their values and identity in concert with others to form a loose contingent of sorts. Specific tenets are often the basis for these food culture decisions and usually involve the exclusion or inclusion of food items based on perceived wholesomeness or toxic properties. Many food culture choices are easily recognizable as large groups of people have embraced them—vegetarian, vegan, locavore, health conscious, ethnic, spiritual, and consumerism are not foreign to our ears, even if we might not understand nuances of the customs.

Even as growing numbers of people subscribe to various food culture themes, adherence to the practices remains solidly at the personal level. There is no registry, membership dues, or accountability. One simply advertises their preferences within their network, and communes with those who share like values. As I noted in Chapters 1 and 5, my partner and I only eat "traceable" meat. In order to manage this, we need to be polite but firm about our commitment when invited to dinner at someone else's home. We can offer to contribute a dish that fits our food values or ask if the meal will be vegetarian or seafood based. It becomes an opportunity to discuss and share our views, raising the awareness of others who may never have thought about how eating commercially grown meat impacts the environment or their health. Some friends consider us vegetarians and move on; others make an effort to offer traceable meats such as venison they have hunted themselves or acquired from a known associate.

Occasionally, discussion of food preferences will expand into a deeper culture defining moment that goes beyond meal choice to the underlying substance of a belief. When my close friends jest that I need to know my meal's name, it shows all of us are in that moment recognizing that what is being eaten was alive. We have rejected the cultural trend inspired by industrial agriculture to consider food as lifeless. That recognition is the first step to respecting that food is life giving. It can inspire a perspective shift from the ego-centric to a community based ideal. As a result, while some practices revolve around notable diet rules, other movements have developed to focus more on access for disadvantaged groups. Concepts such as food security, food justice, and food sovereignty, consider how society's current food culture is not equitable for all members.[5,24,25,26]

I have always been personally aware of issues related to food security as noted at the beginning of this book, but until researching and reading so many diverse perspectives, I had little concept of food sovereignty. Because my own food culture was so undefined when I was young, not anchored in ritual or meaning of any kind it made it harder to comprehend that someone else might have a strong food culture that was restricted beyond their financial means by government policies, environmental destruction, forced relocation to a new climate, or extinction of heritage seeds. While we are all human with the same basic needs, how we fulfill those needs culturally adds complexity to our relationships.

In the United States, many give no thought to their food culture, eating what is familiar or easy to procure and prepare. Lack of consideration of one's diet does not mean they have no food culture though. Awareness or lack of awareness of food culture doesn't change the fact that we all eat to survive and that we tend to eat according to certain preferential norms. The circumstances which cause us to become more aware of our personal food culture may help develop a greater appreciation for the eating experience in some cases, but not all. Perhaps this is because choosing foods as a way to declare one's values is still missing the important component of recognizing food as a source of life energy, given freely by the living being that created it.

Changing of food habits due to trends or popular "alternatives" to define oneself can have distinctly diverse outcomes. The individual might think differently about their food and their identity as an eater in a larger ecological sense that is not human centric, which often nurtures empathy and can be developed into a respect for food sources. Conversely, the individual can wear their new food culture like a lapel button seeking approval from peers while never developing a deeper curiosity about historical food culture, and how their choices may increase food insecurity overall. Decisions about specific food choices may benefit the eater personally, but without considerations regarding the extended impacts of those choices, the end result could be more harmful than helpful to the rest of the living beings sharing the planet.

The years I have spent researching food from a variety of perspectives has led me to conclude that personal food culture is the foundation for all other personal choices regardless of our cognizance. I have noted firsthand many reactions to that concept, very often leaning toward discord or dissent, but I think there is significant evidence to support the idea. Importantly, I have observed that our food choices demonstrate our level of self-love. The desire to reform and upgrade how we feel through what we eat is an act of caring at the most intimate level. Lack of concern or interest over food choices is quite often paired with sentiments that the individual either has little influence in triggering a personal transformation or change at this point is too late to matter.

Even for the most pessimistic, when a meal is presented along with the intent behind the menu item selections as tailored to the guests, it demonstrates the importance of the eater as much as the food to be eaten. I cooked this dish for you because . . . , I was thinking about you when I saw this fresh _____, I remember you telling me about a favorite meal from your childhood that you miss, so I recreated it for you. These phrases contain thoughtfulness, generosity, and love; when someone feels valued and cared for, it makes it easier for them to value and care for themselves, and then to pass it on to another.

Love, like most emotions becomes stronger when we share it. As we begin to recognize ourselves as an incredible combination of physical and emotional being that is dependent on food for survival, we gain the potential to develop both awe and humility for our very existence. Understanding ourselves on a deeper level like this can spark a realization that all living beings are amazing in their own right. Plants provide us ALL the food we eat, sometimes by feeding another being that we then eat, and sometimes directly. They do this by literally creating nutrition from sunlight, water, and minerals in the soil. We can develop an appreciation for the magnitude of this gift by dedicating focused gratitude to our foods for their life-giving power. This may be easier to visualize with whole or minimally processed foods but is a practice worth nurturing at every eating opportunity regardless of the disguises that processing may have bestowed upon a particular meal item. Self-love cultivates feelings of admiration and reverence for our own existence and helps put that in context with the living world around us. Self-love breeds humanity, kindness, and empathy.[11]

Once compassion for our personal being is nurtured by our realization that nutrition is actually given to us as part of a transfer of life energy, we may expand our capacity for generosity to others. We want other humans we care about to feel the positive impacts we have experienced. Honoring and accepting the energy provided by non-human living beings is rewarding and reinforces our gratitude for living in this moment. Sharing food choices, the catalysts that prompted them, and the dividends received from them demonstrate our engagement in the world around us. Our desire to lift up, enhance, and expand our personal network multiplies our own enjoyment through the collective feedback of those who also appreciate the rewards earned.[33] The practice of encouraging others to try choosing foods for the specific nutritional attribute of freshness or wholesomeness that embodies life energy translates beyond the ethereal as well, impacting food access and food culture in tangible ways.

As part of the economy, encouraging others to seek specific food items such as organic or cleanly grown, local, traceable, minimally processed, or low in refined sugars, often helps expand demand and increases the availability in a given area which benefits the community at large.[15] While some argue this can raise the prices of these desired items making them unaffordable for portions of the population,[4] without demand these items simply won't exist in the market at all. Promoting an increase in the number of farmers and food processors who provide the desired food attributes is an effective way to shift food production in the moment toward a more accessible and sustainable system tomorrow.[23,32] Admittedly imperfect as the only solution to food access, it does provide a way for concerned eaters who can afford it to take action in changing our food culture now.

A return to the practice of keeping a garden where fresh food can be grown at reduced cost is yet another way some choose to nurture a healthy food culture, personally and for neighbors and family who share in the harvest.[32,33,132] Developing a direct connection to any part of our diet through patiently caretaking for another living being that gives to us in the form of something that becomes a part of our essence when we ingest it is the ultimate reward. From a simple herb box in a sunny window, to a tomato, pepper, or bean plant in a pot on a porch or balcony, to an area in a yard planted with the eater's favorite foods; any time we remove money from the equation, give of ourselves via time and gratitude, and receive back from a living food source, we increase our appreciation for our place within nature.

Still, not everyone has the resources or knowledgebase to develop such a close connection to their food sources. Access to the smallest plots of land can be completely out of reach for many.[4] Attaining seeds for foods within one's traditional heritage might be impossible as the global industrial agriculture machine ignores the benefit of heirloom varieties for the production of a limited number of monocultures, causing the traditional cultivars to become extinct or increasingly rare.[23,30] Native American's who were stripped of all earthly possessions, persecuted and repeatedly moved to reservations and climates that are entirely different from the lands they had a deep connection to, have no way to recover some seeds for foods that are sacred to them.[28] If they have managed to remain close to their ancestral homelands, the foods they know are often out of reach as a result of ecological damage and hunting restrictions created by our culture;[4,28,29] our culture that simply doesn't recognize food as life and food is life.

The historical injustices of Indigenous farmers all over the world being forced from their land or manipulated into growing industrial agriculture's limited variety of monocultures is well documented. These are the effects of our apathetic food culture giving up and letting industrial agriculture take the reins, and everyone's food security continues to weaken as a result. Others may even find the idea of growing their own food in direct conflict with their personal identity as a result of historical abuses like slavery.[4] Minority farmers face a multi-faceted struggle; as farmers fighting ongoing discrimination for common agricultural resources provided by the government to nonminority farmers[133] (it's disgusting that this issue persists even as we begin the third decade of the twenty-first century), for Black individuals fighting for an identity separate from the slave forced to grow food for others,[4] and for many Black and Brown communities fighting to overcome diet based diseases that disproportionately crush their quality of life.[4,6]

As a gardener, I can't help but feel mournful for anyone who has been forced away from a direct connection with their food sources for any reason, as I know the benefit I feel from having an intimacy with nature and the life energies that flow throughout cultivated spaces. I also understand that the experiences of others are sometimes so different from my own, that their path to a positive relationship with whole foods may be entirely different. There are so many possible ways in which people can engage with food production, all of which impact access and culture both positively and negatively.

While I hope for more people to connect on a deeper level to their food and the life energy it gives because I believe it will enhance our life experience and lift up our culture, the reality remains that many have a more serious survival crisis to overcome first. An increased awareness regarding food culture and choices spreading through communities helps highlight inconsistencies in food availabilities for those who struggle with hunger and malnutrition. Awareness encourages discussions and debates and promotes ideas to be fleshed out and articulated for their priorities; open communication is the key to expanding our comprehension of the cultural differences at play within our society. Demonstrations of projects that take the most food insecure into account show how our private and personal values can be translated and applied in the context of our community if we make the effort to engage, listen, and work together.[33,34,35,132]

Of course the opposite is also true; we can quietly maintain our personal food culture without consideration of whether our neighbor has enough to eat. We can feel relieved of guilt by donating highly processed foods to organizations fighting on the front lines of hunger today and not give a thought to building a better tomorrow while we purchase fresh foods to consume, completely oblivious that our own food security is also at risk. I think this mentality is hard to maintain when one recognizes food is life energy and we are the receivers of another living being's generosity though.

If awareness is cultivated, gratitude and humility follow. The many food "movements" already influencing eaters throughout our country demonstrate how personal perception leads to action. The conflict between the normalized and discriminatory food culture in the United States[4] and individual based values regarding food consumption sets the stage for a dual identity that will need to be reconciled. Influences such as health concerns, financial status, regulatory practices, advertising exposure, religious and/or spiritual beliefs, and our core values, all coalesce to form our defined personal food culture, and separate from that we maintain a sense of what we perceive is appropriate for others. Freedom of choice might act as a defense chant allowing some to feel absolved of culpability for those who struggle with food insecurity. But realistically the concept of freedom of choice completely disregards that "Social, political and economic institutions and policies are enormously influential on people's capabilities with respect to food security."[7]

More and more, food advocates are calling out "freedom of choice" for what it is, a lack of compassion. Others are taking action to bridge the gap between clean fresh food access and financial, racial, and geographical barriers through an ever-growing number of projects.[33,34,35,132] Whatever the individual's approach to food choices, those decisions become ingrained and reinforced in our personal culture every time we take a bite. Proclaiming specific food preferences and sharing them with others can be the birth of resistance to a food culture lacking in empathy that no longer represents us, and realignment between our actions and the stated values we want to achieve. Personal food culture is the most important tool we have to promote change in the world around us.

How does one take ownership of their food culture then? Perhaps the most difficult step is committing to and allowing time for introspection. The fast pace of our technologically based lives keeps us moving constantly. We are always on the run, rarely slowing down to eat with gratitude or think about what we are eating and why. Introspection and self-love need quiet moments of focus to blossom. The body requires time in the moment to provide feedback. Being open to the development of awareness requires acceptance that we may discover miss-steps or conflicts between our values and our actions. Forgiveness is part of practicing self-love and embracing gratitude.[11] For me, thinking about the pure and selfless energy fresh whole foods fill me with when I eat them helped me develop empathy and clarity. For you, the path may be more socially conscious or fed by a dedication to family heritage.

So long as our culture takes note of increasing food insecurity, and works to halt and reverse the historical trends, we can begin to eliminate fear. Fear for the millions of people who are food insecure, fear for those concerned the calamity might worsen, and fear that we might be impacted personally in the future. We must believe in our power to effect change, but faith is a double-edged sword. Faith that others will carry the mantle of revolution, and our contribution is negligible is dangerous. Instead, we must embrace faith that the current trajectory of our food systems is not irreversible, and that projects rooted in ethical values will yield results more quickly than we can imagine when we all actively work together. Self-love, gratitude, and empathy help us redefine the world around us and our place in it. We begin to see that family includes more than just blood relatives. Family is all of the living beings we feel connected to, supported by, and responsible for. We are all related by our need to eat, to absorb nutrition, to feel satiated and content, and to grow into a better version of ourselves. Our food culture is a mixture of fear, faith, and fellowship with family, constantly being stirred and seasoned. To tame the bitterness of fear, we must act with faith to bring out the full flavor of fellowship for every soul.

CHAPTER 8

THE 4 PS: PERCEPTION, PERSPECTIVE, PROACTIVE, PERSISTENCE

As the previous chapters have shown, there are countless ways to look at and consider food. Every individual has a unique conception of food based on myriad combinations of factors. Some recognize the imminent risk of future food insecurity for all because of climate change and environmental destruction, some are living with culturally and race-based structured insecurity as a major impactor in their daily lives right now. For others, personal identity may be intimately tied to food in relation to health or cultural heritage. Many implicitly trust that economic and political systems will always provide nutritional sustenance, so the consumer does not need to be concerned and can simply buy based on taste and convenience preferences with no negative impacts to their well-being.

With such varied opinions, how can a consensus be reached that food should be thought about with more regard to improve each person's enjoyment of life? How do we inspire change on a large scale for the benefit of all? I think both the problems and solutions can be defined by the 4 Ps: Perception, Perspective, Proactive, Persistence.

PERCEPTION.

Increasingly in the past two or three decades, writers have shone a light on the many issues plaguing our food systems; my shelves are over-flowing with books that discuss food from every conceivable angle. Their works have inspired a portion of the population in our country to think somewhat more carefully about what they ingest, and a smaller group to become so passionate about resisting the industrial food complex that alternative food procurement options like farmer's markets, CSAs, and local specialized farms are becoming more familiar and accessible in response to consumer demands. As these options and overall awareness about what they stand for grow, a wider segment of the population is purchasing alternative products over conventional options.

Still, an appetite for regularly eating a clean healthy diet is not the norm. Those who pursue something different from a diet of highly processed foods are *the other. Foodies*, participating in the "*alternative food movement.*" The language defining this group as a minority set apart insinuates that most of our community members are completely comfortable with their processed food choices, and see no connection between health issues, stress management struggles, or family and cultural value discord and their diet. Either that, or the idea of devoting time and energy to changing food habits seems like such an arduous task that the reward could hardly be worth the effort. This suggests to me that the real question should be why do we think we are not valuable enough to eat well? From my research, I believe there are three answers to this question that wind like tangled vines creating a cage around us, strengthening each other, silently restricting our movement and growth, and seducing us into passive acceptance.

The most common refrain I have heard is that one person's actions are inconsequential. This is used as both a personal excuse and as an attack to limit the potential achievements of another who may be poised to demonstrate the opposite. It's like the suggestion to use a floatie to cross the swamp versus navigating the mud and submerged trees. While the specific comments may initially appear to contain a sense of concern and caring for another, the underlying sentiment is of fear that another's success will highlight the lack of achievement of those around them. Believing one's actions to be inconsequential is sometimes a result of cultural norms and past failed attempts to reach for more, but I think there is a deeper influence at work that feeds the inaction monster.

Here is where the other two answers come into focus. Beyond the idea of effectiveness of "actions" is the concept of the value of self. Self-assurance. A confident person believes their actions to be impactful because they can see clear connections between their existence and that of the world around them, while another who questions their place and purpose will struggle to see anything but random occurrences happening to them and holding them in place. Focused inward, one's perceived sense of value as a part of a larger conglomeration of living beings triggers action or inaction in response to daily stimulus. Focused outward, one's sense of self dictates our capacity for empathy.

The greater the sense of self-value as a part of a complex system that one possesses, the more one can recognize and appreciate another's value. I suspect that our culture's obsession with competition as the primary motive for every natural system we observe, the mechanism for measuring compensation in workplace environments, and the preferred medium for entertainment, drives and is driven by an overall frailty of our sense of self-worth. Within a society so dedicated to competing, empathy for and the lifting up of another is perceived as weak; a concept that inevitably conveys a sense of risk should we unexpectedly find ourselves in a position of need, and triggers defensiveness which compounds the competitive trend. As a result, charity is applauded to a degree so long as one's own position is not jeopardized. Plans to increase equality are condemned as forced limitations on those who have more than the average. It feels like something is being taken away because dedicating energy toward mitigating fear and building the façades of success that competition requires is exhausting. It drains our stamina and results in a hollow emotional void.

The lens through which we perceive our own self-worth is the same one used to consider the food that we need to survive. The positive embodiment looks like this: I am important. I play a role in the space I live in; I have an impact on my family members, my personal network, my community, and the environment around me. I seek out foods that enhance my ability to sustain and improve my health and well-being, and I want to share that sense of comfort with those around me. The negative: I am secretly afraid I am not good or strong enough. I do my best to fit into a niche where I am needed so I have security. Few people notice my achievements. I eat foods that are convenient and inexpensive so I can save my resources for more important things that make me feel like I am keeping up with my peers so I might gain their recognition and admiration.

Admitting how we perceive our place and value within the world around us, and how an underlying culture of competition contributes to our self-perception, may be the hardest part of improving one's diet. No one can change how we perceive ourselves except for us, and food silently fortifies our deepest beliefs multiple times a day. If our entire lifetime has been spent reinforcing a self-deprecating internal monolog, we likely don't have the skill set to reverse our perceptions overnight. Importantly, while the way we consider food is a reflection of our self-view, I think food can also be used as a tool to change how we view ourselves.

To start, one must commit to taking the time to seek out fresh whole foods on a regular schedule such as weekly. Use that meal to consider the amazing capacity of plants to create nutrition. I have noted this point in a variety of ways repeatedly throughout this text on purpose. This simply truth is overlooked automatically by most eaters multiple times a day. My repetition is to show how easily we dismiss, ignore, and forget this key point. When we accept and understand that food is life, and our lives are partnered with other living beings in a community much bigger than we usually consider, we gain value and support. Native American professor and biologist Robin Wall Kimmerer phrases it in yet another way, "Plants know how to make food and medicine from light and water, and then they give it away."[134] If your meal includes meat, think about the plants the animal ate during its life, and the sacrifice the living being made to become part of your meal.

Make sure this time is free of distractions. Think about the time spent preparing this meal as a form of worship and gratitude both for the food items that are sharing the nutrition they contain to benefit our bodies and for the incredible processes within our bodies that absorb those life-giving nutrients. Pay close attention from the first smells that trigger our appetite, to the enjoyment of the flavors and textures of the food, and the resulting satisfaction we feel as our hunger fades. Share the experience with others when possible, vocalizing praise for the taste of freshness and the connection between us and the living beings who support us. The repetition of this exercise will reinforce the value of fresh foods and how they impact our bodies.

Visualizing the value of living and life-giving foods being consumed and absorbed into our flesh helps us realize that our bodies now contain the valuable life energy we have eaten. Our brains have been provided the nutrients needed for clear thoughts, our organs have each received what they need to pump, flex, stretch, and work within the various systems of our physiology. **We become the manifestation of the value we have taken in.** We are part of a larger cycle in which we receive energy from whole foods and then use that energy to be an active and engaged participant in the world around us. This message is the best weapon against a fear of being inadequate or inconsequential. We don't have to suddenly believe we have value after years of doubt; we can observe how we physically gain value through the ingestion of plants (and animals) that are our partners in survival and enjoyment on this planet.

PERSPECTIVE.

Perspective is equally as important as perception. Where perception is individualized and internal, the sorting of outside influences and how they relate to us, perspective is the mechanism we use to identify and place ourselves externally within the world. These two concepts are tightly intertwined. The individual who perceives themselves as valuable will be able to clearly articulate their influence on the world around them, capable of identifying how they relate to complex and interactive systems. They will not be easily threatened by another's success; they will encourage and lead others. They will feel a responsibility to help those who suffer, understanding they are not directly at fault, but that they have the power to help lift another up and by doing so improve the world around them making it more secure for all.

The opposite is the individual who lashes out to hide their fear of irrelevance. They crave praise, are boastful in attempt to deflect scrutiny, and believe that everything they have must be protected so it is not taken away. They don't believe their actions reverberate because secretly they feel impotent; they perceive their actions as finite with no momentum to carry on. This individual admires power, money, and influence as evidence of relevance. Competition is the cornerstone of life. Food is only one of many things to be acquired, controlled, a means to an end, and a status symbol.

If we perceive our influence as limited, our perspective about ourselves and others will mirror that view. A perspective rooted in the idea that one has no inherent power as an individual unless they have achieved a certain status will reinforce a self-deprecating internal monolog. The thought process can look like this: Why put so much work into writing a book? Will it really reach a large enough audience to be meaningful? If it does, will enough people make changes in their attitudes toward food to even be measured? If the likelihood of influencing eaters toward a more rewarding, healthy, and sustainable food system that is equitable and raises society to a higher level of enjoyment of life is remote, why sacrifice years of effort on the endeavor? Perception and perspective work together to breed doubt as a defensive measure and that doubt can cast a powerful shadow. Even with a strong sense of self-value, taking action can be daunting, and the impacts of one individual always seem narrow when viewed through a wide lens. It is truly difficult to see the full measure of how our actions impact others well beyond our ability to observe firsthand.

Focusing perspective on a more observable scale, we can increase our awareness of the tremendous influence we have on our local network. Family, friends, co-workers, community members; our personal networks are wider than we can comprehend in most cases. The tip I give a waiter or delivery person has influence beyond our brief exchange by influencing their spending budget. My attitude at work impacts the environment my co-workers must spend time in; I can uplift them, or I can draw attention to all the negatives we must cope with daily. They can feel supported or spend energy to ignore me taking away from their ability to excel, or they can commiserate which results in all of us being stagnate. They then carry my influence with them to their families or to other groups they participate in. Adjusting one's perspective to recognize the amount of sway we do have versus diminishing our significance by only considering it at a larger scale, is an important way to recognize our relevance.

Again, food can be a tool to help revise our mindset. After each meal where you visualize the value being provided by the living beings that support our physical, mental, and emotional wellbeing, spend the next day noticing the energy, clarity, and comfort that follows the meal. Did you sleep better, wake more refreshed? Did your guests or family continue to give positive feedback? Is your attitude towards the new day more positive? Do leftovers eaten at lunch give you a second boost of the perfect energy gift delivered by the plants that first collected sunlight and transformed it into nutrients for your consumption? Considering the extended impacts of the meal can help demonstrate that energy transfer is a never-ending cycle, just like the actions of the individual that do not fade into a vacuum but continue to be passed on. Maintaining a perspective that acknowledges our influence rather than dismisses it supports a perception of self-worth, and believing we have value helps us see the magnitude of our persuasion on the world around us.

PROACTIVE.

As we use fresh whole foods to inspire us to feel better about ourselves and expand our concept of personal influence, I think it becomes easier to feel proactive. Knowing that my actions have an effect on my own well-being as well as those around me is empowering. I feel a responsibility to not squander or be careless with this potential. Here is where empathy is given the best opportunity to blossom. I acknowledge the gift I am given each time I eat, and I see how that energy is molded by my emotions and flows outward to touch all those I encounter. I am the determining factor for whether that pure energy, originally created by a plant being, maintains its uplifting quality or is infused with fear and doubt.

Our perception and perspective strongly influence whether we will be proactive or reactive to our surroundings. The confident individual who embraces their potential to impact others will find it much easier to act with purpose. The individual who is unsure of their place will respond to every event as if there are no connections tying the flow of energy together. As noted previously, while researching this topic I set a goal to host small dinner parties at least twice a month for over two years. I invited a wide variety of people, some who I have a very close relationship with and others I knew only casually. I shared ideas about food and listened to feedback from all. Every single dinner was crucially important to me as I built an understanding beyond my current perspective about how others view food.

Each guest took home with them value in the form of clean healthy nutrition from the food offered to them, my gratitude for their time and contribution to the conversation, and perhaps a wider appreciation about their influence than they may have had when they arrived. I received from them ideas to ponder and viewpoints I hadn't yet considered. Thoughts about the role food plays in their lives might trigger further consideration for them because of the experience, or they may simply have a fun evening to recall later. In any case, everyone who sat at my table for those dinners contributed to what I have written in this book because I proactively engaged with them. A proactive attitude helps sustain awareness, keeping it fresh and alive in the moment.

Being proactive isn't just an optional enhancement to our mindset though, developing a stronger connection to the food we eat and the energy we gain from it requires a commitment to proactivity. The exposure to the consumerism culture that dominates business and government in our country is inescapable. It permeates the sights and sounds we interact with countless times a day. Choosing to disconnect from the outside world and create isolated islands of purity in food experiences takes a degree of proactive effort, but that effort is rewarded exponentially in my opinion, and it becomes easier with practice.

PERSISTENCE.

The last P, persistence, describes a regular proactive practice of connecting with food and feeding ourselves with not only the complex chemical components needed for our bodies to operate at peak performance but also the ethereal energy of consumable sunlight imbued with intangible generosity by the living beings who create what we eat for us. We can choose whether to seek out nutrition that uplifts and infuses us with limitless benefits encapsulated in the perfect form of the fresh fleshiness of plants, or to absently swallow the left over remains of previously living beings who have been churned, ground up, stripped, chemically treated, supplemented, and reconstituted. It's a simple decision between an intensely supportive experience in building a positive perception and perspective of identity, self-worth, and expansive potential, or a repetitive and monotonous negative confirmation of doubt and fear.

Persistence does require effort and a bit of skill to master, and the gains take time to measure accurately. As I have researched and written this book my food habits have changed and improved through trial and error. The transition didn't happen overnight, rather it was a steady trend of small improvements over time. I haphazardly discovered and employed some techniques that I later found described and summarized well in a book called *Atomic Habits*. The author and former professional athlete James Clear notes the importance of raising awareness and making it easy to repeat the habits you want to build.[10]

This book you are currently reading has been my attempt to jump start your thinking about what you eat and why from the start. Yet I have only begun to scratch the surface, I couldn't possibly have raised or answered every question one should consider with regards to their relationship with food. The most important concept that keeps presenting itself throughout every piece of this investigation is that a stronger partnership with food can provide a benefit to every individual, and none of us know the magnitude of what there is for us to learn if we open our minds. Hopefully, I have demonstrated that there is a great deal to think about and many angles to ponder, whether your focus is entirely self-centric or leans more toward community based. Identifying the factors that inspire a personal food reawakening are the most important step in this moment.

When it comes to the actual collection and preparation of food, getting started can be the hardest part, especially if you have never looked outside the walls of your favorite grocery store and the aisles of processed boxed foods. Some communities make it easy to locate alternative sources for the fresh local foods that will give you the greatest reward for your new curiosity. In other locations (both urban and rural), "food desert" conditions are a very real challenge. Lack of local fresh food options can be overcome though, especially if neighbors come together and get creative about how to welcome and compensate fresh food providers looking to support a new community. Passionate leaders are needed to start the conversation and visualize the community's needs and resources; not an easy task, but one that has a plethora of possibilities if imagination is nurtured.

For the individual, two components are necessary: time and skill. Preparing fresh foods requires a dedication of time, but perhaps not as much as you might think. Making every meal from fresh ingredients doesn't mean you must be chained to the kitchen every single day. I try to balance and plan ahead with some easy tricks. I may take an afternoon to make a large pot of chili or soup that becomes a couple of meals over the following days and use an Instant Pot to pressure cook jars of the leftovers (the extra could also be frozen). In the future when I am running short on time, I have a homemade meal from quality ingredients ready to pull out for a meal later. Chili is also an adaptable meal base. A jar of chili makes for a warm lunch on its own and topped with cheese and or a slice of hearty bread can be an even more substantial meal. Take it one step further and add tortillas, eggs, cheese, and maybe avocado and you have a well-rounded meal for a family. I have also never made only one pan of vegetable lasagna. If I am going to do the work and turn the oven on, I make two and freeze the second one in individual or double servings for a later date.

In the late summer, when local tomatoes are at their cheapest and best tasting, I make a huge batch of red sauce using fresh herbs from my garden and preserve it in jars like the soup. Through the winter, I use the sauce to make lasagna, pasta dishes, and chili. The flavors and quality of the fresh ingredients that were locally sourced still carry their special life-giving gifts and provide taste that can never be beat by store bought sauce. The best part is that I also reduce my overall food costs by buying in season as well. You might assume I had a strong culinary background from my affection for fresh whole foods, but I didn't grow up learning how to cook; in fact, I was quite useless in the kitchen for the better part of my adulthood. My ability to learn was entirely governed by my attitude and interest, not by any natural ability carried in my genes. Finding ways to make it easy helped me get started and stick with it, like Clear suggests in *Atomic Habits*.[10]

I have often thought about a way that more people could utilize methods like making large batches of soups and red sauce and cooking seasonally even if they didn't have the space, kitchen implements, skills, or confidence. I would love to see community kitchens become a common concept. I envision a space, perhaps staffed by someone with some cooking experience, where people could go to prepare soups or other foods and get assistance if needed. The community kitchen could source local in season ingredients, provide recipe suggestions, and might be set up on a sliding pay scale to allow wider community member access. Both individuals and groups could schedule cooking times. Getting a group together could help defer costs and be a fun family, social, networking, team building, or even charity driven activity. The kitchen would integrate and promote multiple factors including local sourcing, seasonality, cultural and social connections, strengthening of the local economy, improvement of community health, environmental waste reduction (composting), resource maximization (buying bulk), and convenience. Ideas like this could help inspire a new way to think about fresh food and reject the notion that cooking must be a dreaded chore.

Creating a practice of eating fresh whole foods for some meals, and interspersing meals made in advance from fresh whole foods breaks up the work involved, and the benefit remains. Interacting with food at its freshest, and visualizing the rewards gained through ingestion can be paired with having convenient and fast meal options at the ready. The two concepts do not need to be in conflict as marketing and advertising so often suggest. We simply need to imagine something better than the limited options and choices we are bombarded with by media. Using food as an inspiration and creative spark requires only a conscious shift in attitude and attentiveness to perception and perspective. From there, the possibilities unfold. Kimmerer, Native American scientist and decorated professor, notes,

"(T)he power of ceremony: it marries the mundane to the sacred."[134] Bringing a degree of ceremony to meals is crucial!

I am completely convinced that as we interact with fresh whole foods and become more in tune with the wholesome energy available, that energy itself will enhance our abilities to tap even further into what is waiting for us. Food is necessary for human life on the most basic physical level, but that isn't the end of the story. Food is the key to accessing and embracing all that lies beyond the mechanics of physiology. Life. Our emotional and visceral experience of this moment; the fullest and maximum potential of what is possible. Regardless of what one believes will happen after their human body ceases to exist, or the reason we exist in this moment with conscious awareness of self at all, the reality is we do exist. We are alive, and our life is sustained by living energy provided by food. Food is life. Eat food with awareness, gratitude, and respect, and allow food to fill you with life in a way that nothing else on Earth can.

CHAPTER 9
CSA (COMMUNITY SUPPORTED AGRICULTURE)

A s noted several times in previous chapters, CSAs are one of the many ways eaters have to reduce the distance between the plants that create food for us to eat and our dinner plates. Early in my investigation of food, I was quite unfamiliar with CSAs, and the few people I talked to who were savvier about the topic had vastly different opinions to share. I wanted to better understand CSAs before recommending them in this book so I devoted an entire study to learning all I could about them. I talked to farmers, I interviewed friends and friends of friends, and I read books by some of the pioneers of the CSA movement. I even put together a very basic survey to get a better sense of how my knowledge and awareness (or lack of) compared to the society at large. This is also when I began the regular practice of hosting people for a meal in my gardens and home to explore how others thought about food.

I think there is value in what I discovered even if the focus of this chapter is much narrower that what has been discussed in this book so far. I've included this chapter in the hopes that if the previous chapters have resonated or inspired you, this information may be helpful as you build your new food provisioning practice. If you have had an experience with a CSA in the past and it didn't meet your needs at the time, I especially encourage you to read this chapter as I have found that no CSA is like another. Like clothing or hair style, you have to find the right fit for you and to do that it is helpful to know the breadth of possibilities before closing the door after one disappointing experience. I do ask that you keep in mind that I wrote this chapter when I was first pondering ideas about food. Since then, I have researched extensively and developed the ideas I have shared throughout this book, especially in the immediately preceding chapter. What you will find below is a snapshot of my mindset five years ago.

CSA groups are developing in increasing numbers around the country,[32] partnering local farmers directly with consumers for mutual benefit; still CSAs remain a "best kept secret." Why are CSAs so mysterious and what keeps the majority of people around the country from searching them out as a source of fresh food? Why exactly aren't CSAs more popular? I asked Gordon Sacks of 9 Miles East farm how local growers could better tap into that group of people who would be interested in their products if they just knew more about them. He replied, "How does a farmer know how to reach you?"[111] I think the pool of potential consumers who would enjoy the CSA experience is greater than the number of CSAs available, which should mean that CSA numbers should be exploding to meet demand versus just growing at a steady pace. CSA offerings should be as familiar to consumers as where the grocery store is located and considered common knowledge for a society so focused on health and wellness themes in nearly every facet of our daily lives.[135]

Throughout this study of CSA, I kept returning to the question, why aren't people more motivated to eat fresh healthy food? Could it be that individuals hear the message loud and clear, but don't understand what it means to eat healthy? Many leadership and management classes often start with the concept of being clear about what it is you expect as a manager. Leaders can't use vague terms like "work harder" or "do better"; employees don't know how to translate that into actionable steps. Instead, you must use precise language and set definable goals such as *I expect you to meet your weekly project deadlines or communicate why that isn't possible at least eight hours before our team meets to discuss progress each month.*[136]

Is the message of "eat healthy" too vague for consumers? Have most individuals lost the ability to distinguish between the effect of real fresh food and processed food products on their bodies?[137] Do our busy lives, facilitated by nutrient substitutes available on the go, really allow us to find more happiness than we might have if we took the time to enjoy the complete experience of meals made from scratch?[138] Is the connection between the natural chemical compounds found in organically grown food, and the way it impacts our health so cryptic that only scientists can comprehend the relationship?[139] Is the average consumer so immersed in their own field of specialty that they don't have the time or interest to understand the basics of how the economy works with relation to the production of food; and how they influence the system?[140] Do consumers so trust in the system, that they are willing to leave the topic of food security to the same political leaders that they complain are completely self-serving?[141]

The possibilities of potential reasons for apathy and separation from food are long, and I can only begin to ponder what the majority of our community holds as their truth with relation to fresh food. To propose with any legitimacy the answers to these questions regarding food and our lives, I must find out what the consensus of opinions is in a varied group of people. The only way to accurately examine each of these factors is to ask a series of questions designed to try and uncover what thought processes are prevalent, and if presented with an idea, would the general public take the time to reconsider their own understanding of food. To that end, I developed a survey and distributed it to people in various areas of my network. I also requested that the recipients pass the survey on to others that would be willing to take the time to complete it. I received just over one hundred completed surveys back, with respondents from eight states. I now had a variety of perspectives from different age groups, social and economic standings, and geographic locations from which to determine if there were trends that suggest a general attitude of the majority, or if there was a divide between various groups based on upbringing, affluence, or education. Here are the results of my informal survey as they apply to my wonderings regarding CSAs. For actual questions and multiple-choice answers offered you can find the survey results following the resource list at the end of this book.

To begin, I determined that roughly a third of the overall group had some familiarity with CSAs. That percentage was slightly higher than I expected; however, the number of people from that portion who intend to participate in a CSA in the future fell more in line with my expectation. Only a fifth of the entire group surveyed had intentions of being part of a CSA in the future. Why would 40 percent of the consumers who had working knowledge of a CSA decide that fresh locally grown food wasn't enough of a draw? The two most popular reasons given for not re-joining a CSA group going forward were too much food, and alternative methods for acquiring fresh healthy food.

This rings true for my own recent CSA experience as well. I have my own garden, and I think my tomatoes are some of the best tasting gifts of nature I have ever had the pleasure of consuming. When my CSA offered tomatoes, I was faced with few options. I could leave my portion of tomatoes behind and hope someone else enjoys them, which also reduces the overall value of my CSA share as I would now be getting less for my money. I could take them home and hope I have the time to preserve them along with my own bounty for later use. I could also ask if there were any other items in the weekly share that are abundant enough that I could substitute for my tomato portion. Each of these options might be possible one week, but when tomatoes are offered for several weeks in a row, even considering negotiating a work around proved to be a frustrating mental challenge.

Single person households also have a dilemma. Do you try and find someone to split your share with? Your CSA may offer a half share, but then again, they may not. Variety of items offered by the CSA may also impact the "too much food" complaint. I spoke to one survey participant, who said their CSA offered many root vegetables and no greens like lettuce, kale, or other leafy sustenance. Farmer's Markets and personal gardens do offer more flexibility for variety and volume. Another consideration is that those who appreciate fresh locally grown veggies often have a disdain for waste, more so than someone who doesn't think about where their food comes from on a regular basis. For these people, a half a head of lettuce that wilts is truly regrettable, not just unfortunate.

The fact that many of the CSA familiars also regularly attend Farmer's Markets or maintain personal gardens is logical. Some CSAs require or allow shareholders to contribute through working on the farm. As one gains knowledge and experience with growing their favorite items, the advantage of working your own soil on your own schedule can be more convenient than participating in a CSA. This small group is very focused on the food they eat, and their local community; they have searched out and experienced a CSA along with interacting with other forms of fresh food shopping, finding the best fit for them. They consider themselves well versed in CSAs due to their experience. Unfortunately, what they may not realize is the factor that caused them to choose the Farmer's Market or their garden over the CSA they knew may not accurately represent the experience offered by all CSAs.

The uniqueness of each CSA structure is both a blessing and a curse. A working share CSA versus a simple pick-up or delivery share CSA can result in vastly different experiences that may or may not fit the individual. If CSAs are a somewhat mysterious entity, the nuances between diverse CSAs are a whole other level of comprehension. Additionally, a CSA can challenge the adventurous eater to learn new recipes using new ingredients on a regular basis. Arguably growing your own food is the most intimate level of devotion one can have to their own health as it relates to what they ingest, yet the gardener is then less likely to try new foods. Possibly this is in part because of the volume of what they already have, and also because of their knowledge of how to grow certain items more so than others. Being willing to try new foods can open doors to the amazing flavors that nature, and our dedicated farmers, invest their talents into producing. Still this group of food connoisseurs is only a small portion of the people surveyed. A much larger portion had no reference within their knowledgebase for what a CSA was. What is the likelihood of enticing this segment into the CSA community?

The entire group surveyed answered an additional nineteen questions designed to try and isolate beliefs and identify trends within the assemblage. Would the results reflect the same delineation as CSA participants, suggesting that those who have a certain level of food knowledge are more likely to actively pursue healthier options? Or might I find from the answers pool, that there is a much larger group of food advocates available for CSAs to develop around, forming new alliances and partnerships?

After identifying the divide between CSA veterans and those who are novices to the concept, I gave a basic description of the primary themes within CSAs for my audience. My first three questions focused on what factors are most important to consumers with relation to the portion of their food budgets dedicated to fresh healthy food. Two particular points I wanted to discern was, whether the average shopper was blinded by advertising that constantly reinforces the visual attributes of the specimen above all else, and if the shopper simply picks items based on habit and routine. Although all of the food choice factors noted, including organic certification, location of origin, preparation requirements, and appearance received attention, two other factors proved to hold more importance than the rest. Price and nutritional value were each noted by more than a third of those polled as a priority.

The monetary sacrifice to procure fresh food is the driving force of preference for almost forty percent of the crowd, while nutritional value is the peak of significance for only slightly fewer. These findings beget more questions, rather than provide succinct answers. How does one determine the nutritional value of food when the interest in organic certification did not receive comparable notice? Are diners sitting down to meals that consist of the newest and trendiest vegetable to receive attention by a celebrity or medical professional as the cure all without question?[142] If price is so important, why aren't alternative, local, food growers the most popular stop in town for veggies at prices that aren't padded with fuel costs?[143] I knew I needed to evaluate a few more questions before I could form a theory using these data.

Next, I reviewed the amounts and percentages spent on the weekly grocery trip. This section of the survey reflected that the budgets for groceries, and the percentage of expenditure dedicated to fresh whole foods were quite varied. The largest group spent between $15 and $50 on fresh fruits and vegetables, constituting between 15 to 50 percent of their overall shopping bill. Pairing this with the prior results that forty percent of shoppers are fastidiously checking prices, we can determine that many people are eating fruits and veggies on a consistent basis. For most though, it makes up less than half of their weekly provisions.

To confirm this information, a follow up inquiry regarding number of servings of "whole foods" consumed on a daily basis was posed. The description of what constituted a whole food was provided with an example. These results reaffirmed the findings; the highest percentages of those canvassed, eat between one and five servings of whole foods daily. I could now presume that most people eat some level of fresh food at regular intervals, and they are interested in limiting the financial impact on their wallet. This is fantastic news for CSAs looking to grow! The potential member pool is available, even if they are unaware of the opportunities.

Eagerly moving forward, I devised the next two questions in an effort to determine how open my networked group would be to considering a different perspective in relation to the expense of nutritious delights. First consumers were asked if they considered healthy organic food to be costly; 90 percent confirmed they did. Then, would alternative fresh food sources at economical prices be surprising to them; here the positive response dropped to seventy five percent. Some of the assembly appeared to have a general awareness that there are options besides the average grocery store, where healthy food can be acquired for less. This is likely the faction that has a personal acquaintance with CSAs in some form and shops for fresh food to supplement when their garden or Farmer's Market is out of season. Still overall, this demonstrates that there is room, within the mindset of the general public, to entertain another viewpoint, especially if it will benefit their cash flow.

At this point, I decided to tackle the big ideas. My group was warmed up to the topic of food acquisition, and it was time to find out exactly where the populace stood on the issue of health as it relates to nutrition. When offered a multiple-choice question regarding the link between fresh whole foods, and mental and physical health and ailments, would those surveyed respond that they perceived no connection at all between the two, acknowledge a connection that they ignore, or commit that a connection between health and food drives their dietary choices. I was pleased to find that 60 percent of individuals observed a direct relationship between the fitness of their body and mind with the fuel they consume to survive, and the idea spurs their shopping decisions. A total of 35 percent admitted they believed there to be a connection, even though they allow other factors to influence the provisions that they collect to eat. It seems a likely conclusion that the absence of immediate and direct negative consequences as a result of ingesting processed foods lull the public into a false sense of security that there will be time to improve later, at some magical moment in the future.

Although the percentage was very low, I found it incredibly disturbing that 6 percent of the souls who helped me complete this study realize no connection at all between the nourishment they intake, and the function of the human body that cannot continue to move and contemplate ideas without regular meals.[144] Focusing on the positive, this query demonstrated that a large portion of the overall crowd is primed for introduction to alternative food sources that can provide some of the highest levels of nutrition through locally and sustainably grown vittles. The intermediate group may also be swayed depending on their individual reasons for making the choices they do about their diet.

I would hazard an imprecise guess that this sub faction may find food preparation too time consuming or complicated. Here is a potential target audience for businesses like 9 Miles East farm that provides fresh healthy food, already prepared and delivered to your door, for less than what many of the group are spending overall on groceries weekly. Many in the majority may also find the convenience and cost savings of a prepared meal service to be attractive. Together the two groups, almost the entire survey field, have the potential to be open to healthy food alternatives if somehow personally made aware of the options.

My next question was meant to determine if the congregation would be responsive to supporting food sources closer to home. How much of an impact does each person have on their local economy for all items purchased? Here, the group responded with another potent affirmation that the majority recognizes their transaction influence. In total, 20 percent stated they actively encourage support of the local economy through the power of the purchase, while over 60 percent claimed awareness even though their actions are sometimes swayed by convenience factors. Less than 20 percent combined proclaimed that they doubted their measurable impact in the capitalism machine, or they denied the possibility even existed.

I can't help but wonder if those who are ambivalent about their value within the system are also the same ones who don't recognize the importance of nutrition. The idea of human beings without any conviction of their worth has an air of clingy hopelessness and loneliness. These vagabonds of indifference might actually experience the greatest reward if somehow they could be enticed to join a group learning to work the soil, socialize with those who consider their bodies to be a temple of sorts, and nurture their beings with the clean healthy complex chemical compounds found in whole foods.[138]

Reviewing the last two questions in relation to one another touches very briefly on two considerations that could have a significant impact on the choices made regarding food purchases: the adventurous nature of the eater and the skill level of the food preparer. These particulars could both be the bridle keeping the majority from diving into a world of food focus or simply a path by which to initiate the willing into the realm of the passionate. If CSAs are to rally support, perhaps they need to concentrate on providing supplemental resources like descriptions of and recipes for preparing foods that could be unfamiliar and intimidating to newer members.

When asked if one was disposed to trying new foods versus sticking to what they know, almost 70 percent of the conglomerate said they are adventurous in nature about what ends up on their plate. Just over 60 percent also admitted that a lack of confidence in cooking skills limits their courageous spirit. The votes are in! Most of us want to try new flavors, but we are intimidated by the prospect of preparing them. Yet another win in the column for the general populace who would likely enjoy the experience of a CSA, on their path to expanding their health through greater whole food consumption, provided that regular instructions accompany the share.

Overall, I have found that the potential for a CSA revolution exists within the public, still something must be holding us back as CSAs lack acquaintance to the popular majority. Perhaps culture is the culprit. In this country, we seem to lack food culture, unless apathy can be considered a culture of its own. Accordingly, respondents were tested for cultural virility. Over 80 percent claim to eat at least one meal together as a household per week, while 7 percent are the sole members of their household. We may be a society that lacks balance between commitments and home life,[135] but we are making an effort.

The divide between whether recipients grew up in a home with notable culture, and then strove to continue those or new traditions in their homes today, broke close to even with the pro-tradition half weighing slightly heavier. This was a relief to me as I had begun this quest unsure of the amount of subconscious remembrance most people have for the more wholesome foods that preceded our contemporary blend of processed and modified food creations; those convenient and neat packages that seem to have replaced whole foods as the main component of nutrition. Thankfully,

I was able to add cultural awareness and the potential for nostalgia to the general description of at least half the group, which could be considered a positive factor for their possible inclusion into the world of CSA. The other 45 percent of the group cannot be excluded due to their lack of cultural traditions; in fact, they may be more open to forming new traditions focused specifically on healthy foods. The approach CSA organizers use for each group may differ, but the sheer numbers on either side suggests that a CSA should be prepared for both types of new members.

At this point, I had to address the elephant in the room. It seemed to me that in any group I am a part of, there is always at least one person with a food preference or restriction. CSAs could use this to their advantage by advertising clean healthy food, free of many of the causes of distress for some eaters. When asked about concerns that impact food choices, more than half said they had none. Was the other half vegetarian? Think again, barely more than 10 percent claimed that restriction. The second most popular notable dietary distress was allergies at 25 percent; gluten intolerance was tied with vegetarians as runner up for third place.

The bad news seemed to be that less than half the group was primed to be seduced into trying the CSA way of life because of a nutrition related ailment; the good news was that many people, more than half, could experiment with whatever the CSA has to offer without any special accommodations for food handling or variety. At the same time, with allergies as a large concern, CSAs could begin to educate regarding the ways that foods grown en masse with unsustainable practices such as massive monocultures put us at risk for allergy like reactions to the chemical additives of pesticides and fertilizers, not the actual food, while locally grown clean food helps us to avoid exposure.

At this point, I felt that it was important to gather some more perception-based opinions as a way to round out this inquiry. When faced with the choice, nearly 85 percent of my subject pool stated affirmatively that they believed that the United States is progressing toward a "two tier food system" based on social class. Notably, two individuals from a very food-conscious area near Phoenix, Arizona wrote in comments next to their "no" answers that the situation is improving from where their community stood just a few years ago. I take the time to bring light to this, because it demonstrates that social action in a community can have measurable results.

On the question of a general understanding of what the term "Food Desert" was, I found the divide much closer to even, giving evidence to support the understanding that issues related to limited food access in the larger population are not completely obscured in our busy self-centric perceptions. It is possible that some of the majority confirming the segregation between fresh and processed foods, alongside the income class divide, feel that they may fall on the poorer end of the scale. This is backed by the earlier findings of the dollar amounts and percentages being dedicated to the fresh food isles in the store.

Another big idea for the group to consider; why is fresh food so expensive? Presented with a logical scenario describing the amazingly complicated supply chain circus that facilitates the extensive travel of common food items back and forth around the world, padding the end cost to the buyer while earning profits for corporations, nearly 85 percent expected this was a likely occurrence.[145] Piecing other answers together with this one, I can say with some authority that many people allow the expense of fresh healthy food to limit their consumption, despite the knowledge that it affects their overall health. They have an awareness of their power to send a message through buying patterns and may feel an urge driven by heritage to keep the fire of wholesome meals lit in their current households. Because I have determined these statements to be accurate, and yet CSA participation and the pursuit of fresh healthy food through other local and sustainable methods are only growing at a nominal rate, there has to be a key factor holding back the masses.

It seems one must consider the potential that comprehensive thought processes are so stunted in the general public, that stringing these ideas together into a complete multi-level understanding backed by a committed action plan, is actually beyond the current skill set of many. The processed foods that constitute more than 50 percent of the majority's diets are disrupting and distracting us from the very beliefs and convictions necessary to bring about change. I had one more intense ask. Are we so self-absorbed that although we acknowledge faults in our food pathways, and we feel restricted to varying degrees with our financial and or time availability when collecting the nourishment that sustains us; are we individually too comfortable to change? Now that food security had been considered explicitly through the various survey questions as directly connected to each singular person polled; where did opinions fall within the big picture? The minority seemed to shrink. In total, 3 percent were still convinced that availability and quality of food items, regardless of amount of processing, are completely safe and plentiful.

Somewhat unexpectedly, the majority seemed to break apart here, no longer representing a definable theme. Almost 30 percent still recognized differences in qualities of food items available but believed the only limitation for acquisition was their paycheck. A little over 30 percent felt that climate anomalies and fluctuations were the culprit for any shortages they might recognize in their quest for their weekly supply gathering. The highest percentage, just over 35 percent, which is consistent with our original CSA familiar congregate, held a strong conviction that commercial agriculture, and a focus on high volume production, is propelling us toward a period of uncertainty and shortage for all, regardless of financial means.

More than half of the total study subjects appear to have presented me with the following proclamation. We feel that the country is moving in a direction where our already limited ability to buy fresh food will become even more financially challenging, and although we want to uphold the traditions of our youth, we knowingly ignore opportunities to instigate change because we feel that the greater community is not at risk for food shortages in this moment. The potential for a tragic and catastrophic shift with regards to food availability just isn't as real a threat to us as our day-to-day challenges.

We are too distracted and blinded to prioritize our basic survival needs above the struggles to maintain a semblance of comfort in the immediate moment.

As I evaluated what one hundred plus people generously agreed to help me learn, I felt a discomfort that I had stumbled into a spotlight. I supposed that I could succumb to a notion that mankind, propped up by ego, and firmly rooted in resistance to change unless presented without any option for avoidance, must be left to discover its own truth. But we do learn sometimes; we do push ourselves from within to reach new heights, and that rebirth does not always come at the beckoning of calamitous circumstances.

When ideas are presented without challenge, allowing us to safely consider them without feeling defensive, we step up in some cases. Maybe it is in those times when we motivate ourselves, versus feeling driven by other influences or authorities, that we find the path that truly fits. How many of my survey subjects will be more attentive to the information constantly flowing by them regarding what food is available, what foods can help (or hinder) them as they perform and enjoy their daily activities, and to what degree those foods are in jeopardy of disappearing without action on their part? In total, 70 percent noted that they are interested in learning more and actively evaluating their choices. A total of 30 percent recognized that the information is important but can't commit to changing their habits in the near future.

One individual declared that the estimated nine minutes of their life sacrificed in order to be included within the overall evaluation, received absolutely no benefit from the process. I am grateful that this person bothered to return the survey to me at all as they do represent a portion, a thankfully miniscule portion, of our community who is ambivalent to food and its relation to them. Rather than ignore this group due to its limited size, I believe it is a healthy warning sign that those who do believe fresh healthy food is critical to the enjoyment and growth of the individual, their community, and the greater society, must set an example and actively communicate whenever interest is shown in order to dispel complacency and discourage those who don't value food from spreading their perspective.

My final consensus is that CSAs do have a place in many different social groups; matching the right CSA to the right people is key. All varieties of fresh food purveyors must support one another in guiding the consumer to the choices available as they will all benefit. (Gordon at 9 Miles East farm demonstrates a prime example of this community-minded approach.) The weight of the growth of CSAs falls squarely on the CSA members. Their personal experiences are the best example for others regarding the blessings of fresh food access. CSA leaders particularly, need to be attentive to their membership and proactively try to identify if an alternative CSA style might be more conducive for unsettled members. Assisting members to locate the best CSA experience for them will ultimately increase the positive feedback that is passed on and help to generate more memberships for all CSAs in an area.

There is room to push education and awareness with an overall unthreatening emphasis. There is also room to increase the daily portions of whole foods consumed, as attitudes change with regard to the importance of food. In the beginning of the survey evaluation, I considered the complaint of "too much food" associated with CSA portions. Maybe there isn't too much food; the problem may lie with the volume of fresh food we eat regularly, which based on the average daily servings of most, has plenty of room to grow. By sharing and demonstrating healthy eating lifestyle moments, we can gradually, but effectively, create a positive change. To be most potent and lasting, that change must be delicately offered at regular intervals. Like the tide moving grains of sand, consistency blended with patience, understanding, and curiosity, fresh whole foods can magically create a fluid landscape that always has something new to offer the observer.

CONCLUSION

Realistically, do I think that changing how we view food can actually make us more food secure? Yes. Without hesitation. Do I really think it can help us feel more satisfied and complete in our daily lives? Absolutely. Then why hasn't this spectacular change been inspired prior to this point? Until now we haven't been thinking about food as life. We have been thinking about "eating healthy" (an ambiguous phrase) as something we should do, different from what we want to do. Advertising, marketing, and physical addiction to sugar, additives, and fillers convince us that we want processed foods, tell us that we can't live without the convenience, and assure us that substitute and imitation foods are safe and nutritionally complete. The result is a country full of people who consume everything (not just food) mindlessly, or out of a sense of duty, or from fear of missing out on the next thing that is promised to fill that void that gnaws at their soul. Focusing on food is the solution to waking up to the totality of our overconsumption.

Native American scientist and decorated professor Robin Wall Kimmerer asks it best with this simple question, "(H)ow do we consume in a way that does justice to the lives that we take?"[134] How—not should, because when we don't seek the balance of justice in our consumption, our own value, which is measured by the respect and attention we give to others erodes. Humans are not separate from the ecosystems around us; we are not above them. We are not in control, but our actions do have an impact on the environment and all beings living in it including ourselves. We need to realize we are part of something that involves the exchange of life energy. Food is the easiest way to wake up to this mentality. Activist Bill McKibben when writing about his year eating primarily local food noted, "In my role as eater, I was part of something larger than myself that made sense to me – a community. I felt grounded, connected."[92]

Kimmerer eloquently states, "(T)he need to resolve the inescapable tension between honoring life around us and taking it in order to live is part of being human."[134] She follows with, "Reciprocity helps resolve the moral tension of taking a life by giving in return something of value that sustains the ones who sustain us."[134] This begins with becoming aware of the lives intwined with our own. Our food, clothing, homes, transportation, and entertainment all come directly from or involve disruption or the end of the lives of other living beings. Resolving the tension means acknowledging this fact and nurturing gratitude. This theme is repeated regularly in many Indigenous teachings including in the plethora of Native American sources available. It is also echoed in some spiritual practices and occasionally in religious practices.

In addition to gratitude, we need to be able to envision a future that is different, more in line with balance and justice, and also more fulfilling and rewarding for us. A simple thing to say, and an extraordinarily difficult thing to actually do. As a society, we have lost hope. We tend to either acknowledge something needs to be done but think we can't afford to make even the smallest sacrifices to gain the astonishing rewards waiting for us, or we believe as individuals our actions mean nothing so we should continue to take until there is nothing left and deal with that reality in a future we can't even fathom in this moment. We must overcome these mindsets. We are dooming ourselves, our children, and every other living thing around us because of our lack of vision.

This is going to take imagination. We need to stoke our creative brain to dream about something beyond the consumerism mantra we are bombarded with every waking moment of our lives. We need to stop chasing the idea that the enjoyment of life is always just a couple of steps out of reach in the future and find joy right now. Don't worry, if there is a glorious, easy contentment like heaven or whatever you name what is waiting up ahead when this lifetime ends, it isn't going to disappear simply because we take our time to relish in the journey to get there. Instead, the pleasure we have the potential to experience now is reality and can be drawn out and maximized over the longest period possible, if we focus in the moment and let our actions prepare us to embrace a better future.

Food is simply the easiest way to tap into this larger culture redefining moment. Food allows each of us to embrace our potential as humans to restore balance and justice. Food allows us to feel gratitude for and feel loved by other living beings we share this physical and temporal space with. Food can help eliminate fear when shared, inspire faith when consumed with appreciation, and support fellowship in whatever your ideal version of community turns out to be.

NOTES

1. World Health Organization. Health Issues: Nutrition: WHO response. WHO.int; c2021 https://www.who.int/news-room/fact-sheets/detail/malnutrition

2. Blake JS. Nutrition & You. Hoboken: Pearson Education, Inc.; 2020.

3. FAO.org. The State of Food Security and Nutrition in the World. FAO.org; c2021. https://www.fao.org/3/cb4474en/online/cb4474en.html

4. Alkon AH, Agyeman J. Cultivating Food Justice: Race, Class, and Sustainability. Cambridge: The MIT Press; 2011.

5. Smith II, BJ. Food Justice, intersectional agriculture, and the triple food movement. J Agriculture Human Values. 2019;36:825-835. DOI: https://doi.org/10.1007/s10460-019-09945-y

6. Wilkinson R, Pickett K. The Spirit Level. New York City: Bloomsbury Press; 2010.

7. Sandler RL. Food Ethics: The Basics. New York: Routledge; 2015.

8. Nestle M. Food Politics: How the Food Industry Influences Nutrition and Health. Berkeley: University of California Press; 2013.

9. Guptill AE, Copelton DA, Lucal B. Food & Society: Principles and Paradoxes. Malden: Polity Press; 2013.

10. Clear J. Atomic Habits. New York: Avery; 2018.

11. Roth G. Women Food and God; An Unexpected Path to Almost Everything. New York: Simon and Schuster; 2010.

12. Gagliano M, Ryan JC, Vieira P. The Language of Plants: Science, Philosophy, Literature. Minneapolis: University of Minnesota Press; 2017.

13. Ma Rhea, Z. Towards an Indigenist, Gaian pedagogy of food: Deimperializing foodScapes in the classroom. J Envir Edu. 2018;49(2):103-116. DOI: 10.1080/00958964.2017.1417220. https://search-ebscohostcom.library.esc.edu/login. aspx?direct=true&db=eih&AN=128375897&site=ehost-live

14. Dictionary.com. Sustainability. Dictionary.com, LLC; c2021. https:// www.dictionary.com/browse/sustainability

15. Nestle M. What to Eat. New York (NY): North Point Press; 2006.

16. Selhub EM, Logan AC. Your Brain on Nature: The Science of Nature's Influence on Your Health, Happiness, and Vitality. Toronto: Collins Press; 2012.

17. Williams F. The Nature Fix: Why Nature Makes Us Happier, Healthier, and More Creative. New York: W. W. Norton & Company; 2017.

18. Dictionary.com. Organic. Dictionary.com, LLC; c2021. https:// www.dictionary.com/browse/organic

19. Dictionary.com. Health. Dictionary.com, LLC; c2021. https://www. dictionary.com/browse/health

20. Raven G. The Magick of Food; Rituals, Offerings, and Why We Eat Together. Woodbury: Llewellyn Publications; 2020.

21. Kingsolver B, Hopp SL, Kingsolver C. Animal, Vegetable, Miracle. New York: HarperCollins Publishers; 2007

22. Baldwin J. What Ought I to Eat? Toward an Ethical Biospheric Political Economy. J Envir Ethics. Fall2013; 35 (3):333-347. DOI:10.1080/00958964.2017.141722 0. https://search-ebscohostcom.library.esc.edu/login. aspx?direct=true&db=eih&AN=128375897&site=ehost-live

23. Pollan M. The Ominvore's Dilemma. New York: Penguin Books; 2016.

24. Sherrange D. Food Sovereignty and the Future of Regenerative Farming. Weave News. 2 September 2020. https://www.weavenews. org/stories/2020/09/02/2020-9-2-food-sovereignty-and-the-future- of-regenerative-farming

25. Hoover, E. "You Can't Say You're Sovereign if You Can't Feed Yourself": Defining and Enacting Food Sovereignty in American Indian Community Gardening. J Amer Indian Cult Research. 2017;41(3). DOI:10.17953/aicrj.41.3.hoover

26. Gupta-Carlson H. Shifting Ground: Farming, Land Use, and Food Sovereignty. Weave News. 16 October 2020. https://www.weave- news.org/stories/2020/10/16/2020-10-16-shifting-ground-farming- land-use-and-food-sovereignty

27. Sherrange D. On Settler Colonialism: Hearing from the Kanien:keha'ka (Mohawk) Nation. Weave News. 19 August 2020. https://www.weavenews.org/stories/2020/08/19/2020-8-19-on-set- tler-colonialism-hearing-from-the-kanienkehaka-mohawk-nation

28. Salmon, E. Eating The Landscape. Tucson: The University of Arizona Press; 2012.

29. Geniusz WM. Our Knowledge Is Not Primitive. Syracuse: Syracuse University Press; 2009.

30. Nabhan G. Where Our Food Comes From. Washington: Island Press; 2009.

31. Chriest A, Niles M. The role of community social capital for food security following an extreme weather event. J Rural Stud. 2018;64(Nov): 80-90. https://web-b-ebscohostcom.library.esc.edu/ehost/detail/detail?vid=2&sid=3b1ed4ad-12e5-42bb-b6d2-ec74 9ed08093%40pdc-vsessmgr05&bdata=JnNpdGU9ZWhvc3Qtb-Gl2ZQ%3d%3d#AN=132869658&db=ssf

32. Henderson E, Van En R. Sharing The Harvest. White River Junction: Chelsea Green Publishing Company; 2007.

33. Toensmeier E, Bates E. Paradise Lot: Two Plant Geeks, One-Tenth of an Acre and the Making of an Edible Garden Oasis in the City. White River Junction: Chelsea Green Publishing; 2013.

34. Carlisle, L. Lentil Underground. New York: Avery; 2015.

35. Cockrall-King, J. Food and the City. Amherst: Prometheus Books; 2012.

36. Vandermeer JH. The Ecology of Agroecosystems. Sudbury: Jones and Bartlett Publishers; 2011.

37. Pollan M. In Defense of Food: An Eater's Manifesto. New York: Penguin Books; 2009.

38. Meadow AM. Alternative food systems at ground level: the Fairbanks community garden. J Eco Anthro. 2013;16(1):76-84. https://web-b-ebscohost-com.library.esc.edu/ehost/detail/detail?vid=2&sid=17a4edcc-bc63-4754-b54a-f8d8d3ce6511%40p-dc-v-sessmgr03&bdata=JnNpdGU9ZWhvc3QtbGl2ZQ%3d%3d#d-b=ssf&AN=94308273

39. Buchanan D. Taste, memory. White River Junction: Chelsea Green Publishing; 2012.

40. Johannesen J. Shifting consciousness: one student's awakening to the truth behind our industrial food system. J Human & Societ. 2012;36(2):186-189. https://web-a-ebscohost-com.library.esc.edu/ehost/detail/detail?vid=2&sid=67d6ceb8-3fa7-4e71-a26c-bca92 1a5c618%40sdc-v-sessmgr06&bdata=JnNpdGU9ZWhvc3Qtb-Gl2ZQ%3d%3d#AN=77672358&db=ssf

41. Conkin PK. A Revolution Down on the Farm; The Transformation of American Agriculture Since 1929. Lexington: University Press of Kentucky; 2009.

42. Paarlberg R. Food Politics: What Everyone Needs to Know. New York: Oxford University Press; 2013

43. Capon B. Botany for Gardners, Third Addition. Portland: Timber Press; 2010.

44. Berry W. The Unsettling of America: Culture and Agriculture. First counterpoint edition. Berkeley: Counterpoint Press; 2015.

45. Pollan M. The Botany of Desire. New York: Random House Publishing Group; 2002.

46. McIntyre L, Patterson PB, Mah CL. The application of 'valence' to the idea of household food insecurity in Canada. Social Sci Med. 2019;220(Jan):176-183. https://web-b-ebscohost-com.library.esc.edu/ehost/detail/detail?vid=4&sid=3b1ed4ad-12e5-42bb-b6d2-ec749ed08093%40pdc-v-sessmgr05&bdata=JnNpdGU9ZWhvc3Qtb-Gl2ZQ%3d%3d#AN=133623268&db=ssf

47. Brinkley C. Visualizing the social and geographical embeddedness of local food systems. J Rural Stud. 2017;54(Jul):314-325. https://web-b-ebscohost-com.library.esc.edu/ehost/detail/detail?vid=4&sid=f02d7004-7f37-4643-b4f7-4e21eec-c20e9%40sessionmgr103&bdata=JnNpdGU9ZWhvc3Qtb-Gl2ZQ%3d%3d#AN=124938045&db=ssf

48. Ruse M. The Gaia Hypothesis: Science on a Pagan Planet. Chicago: The University of Chicago Press; 2013.

49. Neff RA, Parker CL, Kirschenmann FL. Peak oil, food systems, and public health. A J Pub Health. 2011;101(9):1587-1597. https://web-a-ebscohost-com.library.esc.edu/ehost/detail/detail?vid=6&sid=67d6ceb8-3fa7-4e71-a26c-bca921a5c618%40sdc-v-sessmgr06&bdata=JnNpdGU9ZWhvc3Qtb-Gl2ZQ%3d%3d#AN=508485823&db=ssf

50. Mumber M, Reed H. Sustainable Wellness: An Integrative Approach to Transform Your Mind, Body, and Spirit. Pompton Plains: The Career Press, Inc.; 2012.

51. Banerjee D, Hysjulien LV. Understanding food disasters and food traumas in the global food system: a conceptual framework. J Rural Stud. 2018;61(July):155-161. https://web-b-ebscohost-com.library. esc.edu/ehost/detail/detail?vid=4&sid=65382b79-ee4a-4016-b032- 51795e7d4bfc%40sessionmgr102&bdata=JnNpdGU9ZWhvc3Qtb- Gl2ZQ%3d%3d#AN=130542686&db=ssf

52. Hanh TN, Cheung L. Savor; Mindful Eating, Mindful Life. New York: HarperOne; 2011.

53. Goddeeris L. Food for thought: how and why local governments support local food systems. J Pub Manage. 2016;98(11):27- 36. https://web-a-ebscohost-com.library.esc.edu/ehost/ detail/detail?vid=3&sid=67d6ceb8-3fa7-4e71-a26c-bca921 a5c618%40sdc-v-sessmgr06&bdata=JnNpdGU9ZWhvc3Qtb- Gl2ZQ%3d%3d#AN=119547235&db=ssf

54. Montefrio MJF, Johnson AT. Politics in participatory guar- antee systems for organic food production. J Rural Stud. 2019;65(Jan):1-11. https://web-b-ebscohost-com.library.esc.edu/ ehost/detail/detail?vid=3&sid=65382b79-ee4a-4016-b032-517 95e7d4bfc%40sessionmgr102&bdata=JnNpdGU9ZWhvc3Qtb- Gl2ZQ%3d%3d#AN=134423466&db=ssf

55. Cleveland DA, Müller NM, Tranovich AC, Mazaroli DN, Hinson K. Local food hubs for alternative food systems: a case study from Santa Barbara County, California. J Rural Stud. 2014;35(Jul):26-36. https://web-b-ebscohost-com.library.esc.edu/ ehost/detail/detail?vid=7&sid=17a4edcc-bc63-4754-b54a-f8d- 8d3ce6511%40pdc-v-sessmgr03&bdata=JnNpdGU9ZWhvc3Qtb- Gl2ZQ%3d%3d#AN=97080761&db=ssf

56. Colloredo-Mansfeld R, Tewari M, Williams J, Holland DC, Steen A, Wilson AB. Communities, supermarkets, and local food: map- ping connections and obstacles in food system work in North

Carolina. Human Org. 2014;73(3):247-257. https://web-b-ebsco-host-com.library.esc.edu/ehost/detail/detail?vid=5&sid=17a4ed-cc-bc63-4754-b54a-f8d8d3ce6511%40pdc-v-sessmgr03&bdata=Jn-NpdGU9ZWhvc3QtbGl2ZQ%3d%3d#AN=98041354&db=ssf

57. Inman P. Regional food systems as engines for sustainable econ-omies: how do universities engage? Social Alt. 2015;34(2):39-46. https://web-b-ebscohost-com.library.esc.edu/ehost/detail/detail?vid=3&sid=f02d7004-7f37-4643-b4f7-e21eec-c20e9%40sessionmgr103&bdata=JnNpdGU9ZWhvc3Qtb-Gl2ZQ%3d%3d#AN=109999467&db=ssf

58. Robbins J. The Food Revolution; How Your Diet Can Help Save Your Life and Our World. San Francisco: Conari Press; 2001.

59. Brandhorst TT, Klein BS. Uncertainty surrounding the mechanism and safety of the post-harvest fungicide fludioxonil. J Food & Chem Tox. 2019;123(Jan):561-565. https://web-b-ebscohost-com.library.esc.edu/ehost/detail/detail?vid=13&sid=6dfb734e-2a43-4f9d-bc5c-f4dbd6dd240d%40pdc-v-sessmgr02&bdata=JnNpdGU9ZWhvc-3QtbGl2ZQ%3d%3d#AN=133750724&db=eih

60. Hoy MK, Clemens JC, Martin CL, Moshfegh AJ. Fruit and Vegetable Consumption of US Adults by Level of Variety, What We Eat in America, NHANES 2013–2016. Nutritional Epid Pub Health, Current Developments in Nutrition. 4 Feb 2020.

61. Centers for Disease Control and Prevention: Overweight & Obesity Data & Statistics. U.S. Department of Health & Human Services; 2019. https://www.cdc.gov/obesity/data/index.html

62. Dictionary.com. Balance. Dictionary.com, LLC; c2021. https://www.dictionary.com/browse/balance

63. Foodinsight.org. 2020 Food & Health Survey. International Food Information Council. 9 June 2020. https://foodinsight.org/2020-food-and-health-survey/

64. Bittman M, Katz, DL. The Last Conversation You'll Ever Need to Have About Eating Right. Grub Street. 2018. http://www.grubstreet.

com/2018/03/ultimate-conversation-on-healthy-eating-and-nutrition.html

65. Eatright.org. Dietary Guidelines and MyPlate. Academy of Nutrition and Dietetics; c2020. https://www.eatright.org/food/nutrition/dietary-guidelines-and-myplate

66. Eatright.org. Fitness. Academy of Nutrition and Dietetics; c2020. https://www.eatright.org/fitness

67. Van Den Houwe K, Evrard C, Van Loco J, Lynen F, Van Hoeck E. Use of Tenax® films to demonstrate the migration of chemical contaminants from cardboard into dry food. Taylor Francis Grp Food Add and Contam Part A. 2017;34(7): 1261-1269. DOI: 10.1080/19440049.2017.1326067 http://library.esc.edu/login?url=https://search.ebscohost.com/login.aspx?direct=true&db=eih&AN=124613767&site=ehost-live

68. Carson R. Silent Spring. Boston: Mariner Books; 1962.

69. Buchmann S, Nabhan GP. The Forgotten Pollinators. Washington: Island Press; 1996.

70. Hoefkens C, Vandekinderen I, De Meulenaer B, Devlieghere F, Baert K, Sioen I, De Henauw S, Verbeke W, Van Camp J. A literature-based comparison of nutrient and contaminant contents between organic and conventional vegetables and potatoes. Brit Food J. 2009;111(10):1078-1097. www.emeraldinsight.com/0007-070X.htm

71. Uçurum ÖH, Variş S, Can Alpsoy H, Keskin M. A comparative study on chemical composition of organic versus conventional fresh and frozen tomatoes. J Food Process Preserv. 2019;43e:13964. DOI: 10.1111/jfpp.13964

72. Goldsmith JR, Herishanu Y, Abarbanel JM, Weinbaum Z, Gurion B, Sheva B. Clustering of Parkinson's disease points to environmental etiology. Archive Environ Health. 1990; 45(2):88-94.

73. De Felice A, Scattoni ML, Ricceri L, Calamandrei G. Prenatal exposure to a common organophosphate insecticide delays motor

development in a mouse model of idiopathic autism. J PLoS One. 2015; 10(3). e0121663. doi:10.1371/journal.pone.0121663

74. Sheweita SA. Carcinogen-metabolizing enzymes and insecticides. J Environ Sci Health. 2004;B39(5-6):805-818. DOI: 10.1081/ LESB-200030877

75. Barnhill A, Palmer A, Weston CM, Brownell KD, Clancy K, Economos CD, Gittelsohn J, Hammond RA, Shiriki K, Bennett WL. Grappling with complex food systems to reduce obesity: a US public health challenge. Pub Heal R. 2018;133(sup): 44S-53S. https://webb-ebscohost-com.library.esc.edu/ehost/ detail/detail?vid=2&sid=71d45e84-68f1-4c80-a047-0a17636d-3f94%40sessionmgr120&bdata=JnNpdGU9ZWhvc3Qtb-Gl2ZQ%3d%3d#AN=133008975&db=ssf

76. Smyth S, Entine J, MacDonald R, Ryan C, Wulster-Radcliffe M. The Importance of Communicating Empirically Based Science for Society. CAST Commentary. September 2020;QTA2020-5. https:// www.cast-science.org/publication/the-importance-of-communicat-ing-empirically-based-science-for-society/

77. Smyth SJ. The human health benefits from GM crops. Plant Biotechnology Journal. 2020;18:887-888. doi: 10.1111/pbi.13261

78. Leopold A. A Sand County Almanac. New York: Oxford University Press, Inc; 1949.

79. Prévot AC, Cheval H, Raymond R, Cosquer A. Routine experiences of nature in cities can increase personal commitment toward biodiversity conservation. BC. 2018:226:1-8. https://web-b-ebsco-host-com.library.esc.edu/ehost/detail/detail?vid=2&sid=1da7842f -4e0e-4b4f-acec-2ce15941dc36%40sessionmgr104&bdata=JnNpd-GU9ZWhvc3QtbGl2ZQ%3d%3d#AN=131628493&db=eih

80. Gagliano M. Thus Spoke the Plant. Berkeley: North Atlantic Books; 2018.

81. Kuhn, TS. The Structure of Scientific Revolutions. Chicago: The University of Chicago Press; 2012.

82. Linda Jones PhD, discussion, March 8, 2021

83. Haskell DG. The Forest Unseen; A Year's Watch in Nature. New York: Penguin Books; 2013.

84. Brady NC, Weil RR. Elements of the Nature and Properties of Soil, Third Edition. Upper Saddle River: Prentice Hall; 2010.

85. Wohlleben P. The Hidden Life of Trees. Vancouver: Greystone Books; 2015.

86. Tallamy D. Bringing Nature Home. Oregon: Timber Press; 2007.

87. Lovelock J. Homage to Gaia. Oxford: Oxford University Press; 2000.

88. Moody D. Seven miconceptions regarding the Gaia hyposthesis. Climatic Change. Jul2012;113(2):277-284. DOI: 10.1007/s10584-011-0382-4. http://library.esc.edu/login?url=https://search.ebscohost.com/login.aspx?direct=true&db=eih&AN=76373926&site=ehost-live

89. Oskin B. Sound Garden: Can Plants Actually Talk and Hear? Live Sci. March 2013 https://www.livescience.com/27802-plants-trees-talk-with-sound.html

90. Eatright.org. Eat right this year. Academy of Nutrition and Dietetics; c2020. https://www.eatright.org/health/lifestyle/culture-and-traditions/eat-right-this-year

91. Shuman MH. The Small-mart Revolution: How Local Businesses are Beating the Global Competition. San Francisco: Berrett-Koehler Publishers, Inc; 2006.

92. McKibben B. Oil and Honey: The Education of an Unlikely Activist. New York: St. Martin's Press; 2014.

93. Béné C, Oosterveer P, Lamotte L, Brouwer ID, de Haan S, Prager SD, Talsma EF, Khoury CK. When food systems meet sustainability – current narratives and implications for actions. World Dev. 2019;113(Jan):116-130. https://web-b-ebscohostcom.library.esc.edu/ehost/detail/detail?vid=2&sid=1c681d20-bcb6-47d7-986cac-82cdbbb675%40sessionmgr120&bdata=JnNpdGU9ZWhvc3Qtb-Gl2ZQ%3d%3d#AN=132512136&db=ssf

94. Halbe J, Adamowski J. Modeling sustainability visions: a case study of multi-scale food systems in Southwestern Ontario. J Env Manage. 2019;231(Feb):1028-1047. https://web-b-ebscohost-com. library.esc.edu/ehost/detail/detail?vid=6&sid=3b1ed4ad-12e5-42bb -b6d2-ec749ed08093%40pdc-vsessmgr05&bdata=JnNpdGU9ZWh-vc3QtbGl2ZQ%3d%3d#AN=133750064&db=ssf

95. Benyus J. Biomimicry. New York: HarperCollins Publishers Inc; 1998.

96. Mangnus AC, Vervoort JM, McGreevy SR, Ota K, Rupprecht CDD, Oga M, Kobayashi M. New pathways for governing food system transformations: a pluralistic practice-based futures approach using visioning, back-casting, and serious gaming. Eco & Soc. Dec2019;24(4):1-20. DOI:10.5751/ES-11014-240402. http:// library.esc.edu/login?url=https://search.ebscohost.com/login. aspx?direct=true&db=eih&AN=141296708&site=ehost-live

97. Liu Y, Yue L, Wang Z, Xing B. Processes and mechanisms of photosynthesis augmented by engineered nanomaterials. Envir Chem. 2019;16(6):430-445. DOI: 10.1071/EN19046. http:// library.esc.edu/login?url=https://search.ebscohost.com/login. aspx?direct=true&db=eih&AN=138695536&site=ehost-live

98. Mattick C. Cellular agriculture: The coming revolution in food production. Bulletin Atomic Sci. Jan2018;74(1):32-35. DOI:10.1080/00963402.2017.1413059. http://library. esc.edu/login?url=https://search.ebscohost.com/login. aspx?direct=true&db=eih&AN=127161737&site=ehost-live

99. Hersh B, Mirkouei A, Sessions J, Rezaie B, You Y. A review and future directions on enhancing sustainability bene-fits across food energy water systems the potential role of biochar derived products. AIMS Envir. 2019;6(5):379-416. DOI: 10.3934/environsci.2019.5.379. Pg 385 http://library. esc.edu/login?url=https://search.ebscohost.com/login. aspx?direct=true&db=eih&AN=139625388&site=ehost-live

100. Food Futures Accelerator Series, Skidmore College formal discussion presentation, March 4, 2019

101. Holmgren D. Permaculture: Principles and Pathways Beyond Sustainability. Hepburn: Holmgren Design Services; 2014.

102. Knickel K, Redman M, Darnhofer I, Ashkenazy A, Calvão Chebach T, Šūmane S, Tisenkopfs T, Zemeckis R, Atkociuniene V, Rivera M, Strauss A, Kristensen LS, Schiller S, Koopmans ME, Rogge E. Between aspirations and reality: making farming, food systems and rural areas more resilient, sustainable and equitable. J Rural Stud. 2018;59(Apr):197-210. https://web-b-ebscohostcom.library.esc.edu/ehost/detail/detail?vid=5&sid=65382b79-ee4a-4016-b032-51795e7d4bfc%40sessionmgr102&bdata=JnNpdGU9ZWhvc3QtbGl2ZQ%3d%3d#AN=128564245&db=ssf

103. Zeller BE, Dallam MW, Neilson RL, Rubel NL. Religion, Food and Eating in North America. New York: Columbia University Press; 2014.

104. Tinker G. The Stones Shall Cry Out: Consciousness, Rocks, and Indians. Project Muse Wicazo Sa Review. 2004Fall;19(2):105-125: https://moodle.esc.edu/pluginfile.php/1793005/mod_resource/content/3/M3%20stones_shall_Cry.pdf

105. LaMattina J. Pharma R&D Investments Moderating, But Still High. Forbes. June 2018. https://www.forbes.com/sites/johnlamattina/2018/06/12/pharma-rd-investments-moderating-but-still-high/#1b11bd5f6bc2

106. Varinsky D. The $66 billion Bayer-Monsanto merger just got a major green light – but farmers are terrified. Bus Insider. May 2018. https://www.businessinsider.com/bayer-monsanto-merger-has-farmers-worried-2018-4

107. Dictionary.com. Politics. Dictionary.com, LLC; c2021. https://www.dictionary.com/browse/politics

108. Dictionary.com. Economics. Dictionary.com, LLC; c2021. https://www.dictionary.com/browse/economics

109. Klein N. This Changes Everything: Capitalism vs the Climate. New York: Simon & Schuster Paperbacks; 2014.

110. McKibben B. Deep Economy: The Wealth of Communities and the Durable Future. New York: St. Martin's Press; 2007.

111. Sacks, G. CSA Case Study: 9 Miles East Farm. 2017 (November 5). (V. Kovarovic, Interviewer)

112. Salvador R, Bittman M. Goodbye, U.S.D.A., Hello, Department of Food and Well-Being. NY Times. 3 Dec 2020; Opinion [accessed 2020 Dec 4] https://www.nytimes.com/2020/12/03/opinion/usda-agriculture-secretary-biden.html

113. Healthy Lifestyle: Stress Management. Chronic stress puts your health at risk. Mayo Foundation for Medical Education and Research (MFMER); c1998-2020 [accessed 2020 Nov 14]. https://www.mayoclinic.org/healthy-lifestyle/stress-management/in-depth/stress/art-20046037

114. Dictionary.com. Religion. Dictionary.com, LLC; c2021. https://www.dictionary.com/browse/religion

115. Dictionary.com. Spiritual. Dictionary.com, LLC; c2021. https://www.dictionary.com/browse/spiritual

116. Perry LE. Sacramental Cocoa and Other Stories from the Parish of the Poor. Louisville: Westminster John Knox Press; 1995.

117. Ehrlich E. Miriam's Kitchen. New York: Penguin Books; 1998

118. Sack, Daniel. Whitebread Protestants; Food and Religion in American Culture. New York: Palgrave; 2001.

119. Judith Ehrenshaft, interview, March 5, 2021 (V. Kovarovic, Interviewer)

120. Cindy Castle, interview, March 23, 2021 (V. Kovarovic, Interviewer)

121. Hanh TN. The Fourteen Precepts of Engaged Buddhism. Lion's Roar. Apr2017 https://www.lionsroar.com/the-fourteen-precepts-of-engaged-buddhism/

122. Dictionary.com. Faith. Dictionary.com, LLC; c2021. https://www.dictionary.com/browse/faith

123. Dictionary.com. Morals. Dictionary.com, LLC; c2021. https://www.dictionary.com/browse/morals

124. Stone CD. Should Trees Have Standing? Law, Morality, and the Environment. Oxford: Oxford University Press; 2010.

125. Haskell DG. The Songs of Trees. New York: Viking; 2017.

126. Marder M. Plant-Thinking; A Philosophy of Vegetal Life. New York: Columbia University Press; 2013.

127. Montgomery P. Plant Spirit Healing; A Guide to Working with Plant Consciousness. Rochester: Bear & Company; 2008.

128. Ronfinley.com Ron Finley Project. Ron Finely Project; c2021. https://ronfinley.com/

129. Dictionary.com. Culture. Dictionary.com, LLC; c2021. https://www.dictionary.com/browse/culture

130. Hake M, Dewey A, Engelhard E, Strayer M, Dawes S, Summerfelt T, Gundersen C. The Impact of the Coronavirus on Food Insecurity in 2020 & 2021. Feeding America; March 2021. https://www.feedingamerica.org/sites/default/files/2021-03/National%20Projections%20Brief_3.9.2021_0.pdf

131. Dictionary.com. Ethics. Dictionary.com, LLC; c2021. https://www.dictionary.com/browse/ethics

132. Penniman, L. Farming While Black: Soul Fire Farm's Practical Guide to Liberation on the Land. White River Junction: Chelsea Green Publishing; 2018.

133. Hurt E. The USDA Is Set To Give Black Farmers Debt Relief. They've Heard That One Before. npr.org. 4 June 2021. https://www.npr.org/2021/06/04/1003313657/the-usda-is-set-to-give-black-farmers-debt-relief-theyve-heard-that-one-before

134. Kimmerer RW. Braiding Sweetgrass: Indigenous Wisdom, Scientific Knowledge, and the Teaching of Plants. Minneapolis. Milkweed Editions; 2013.

135. Mental Health America. Work Life Balance. Mental Health America. 2017. http://www.mentalhealthamerica.net/work-life-balance

136. Bodell, L. Kill The Company. Brookline: Bibliomotion, Inc.; 2012.

137. Rose, D. Processed Food Versus Real Food: Why Nutritional Science is So Confusing. Summer Tomato. 2015 (September 1). http://www.summertomato.com/processed-food-vs-real-food

138. Blaszczak-Boxe, A. Eating More Fruits & Veggies May Make You Happier. Live Science. 2016 (July 14). https://www.livescience.com/55407-eating-more-fruits-veggies-linked-with-life-satisfaction.html

139. Gunnars, K. Nine Ways That Processed Foods are Harming People. Medical News Today. 2017 (August 1). https://www.medicalnewstoday.com/articles/318630.php

140. Ackerman, E. Unit One Review Financial/Business Math. Quizlet. 2015. https://quizlet.com/89702709/unit-one-review-financialbusiness-math-flash-cards/

141. Jones, J. M. Democratic Party Image Dips, GOP Ratings Stable. GALLUP News. 2017 (May 16). http://news.gallup.com/poll/210725/democratic-party-image-dips-gop-ratings-stable.aspx

142. Williams, S. Consumers' Growing Appetite for Superfoods. T1 2016 MPK732 Marketing Management (Cluster A) Deakin Business School. 2016 (April 2016). https://mpk732t12016clustera.wordpress.com/2016/04/02/consumers-growing-appetite-for-superfoods/

143. United States Department of Agriculture Economic Research Service. Ag and Food Statistics: Charting the Essentials, Food Prices and Spending. Retrieved from United States Department of Agriculture Economic Research Service. 2017 (September

15). https://www.ers.usda.gov/data-products/ag-and-food-statistics-charting-the-essentials/food-prices-and-spending.aspx

144. Barnes, Z. This Is What Happens To Your Body When You Skip Meals. SELF. 2016; (September 17). https://www.self.com/story/what-happens-to-your-body-when-you-skip-meals

145. Hopp, S. L. The Global Equation. In B. K. Kingsolver, Animal, Vegetable, Miracle (pp. 66-67). New York: HarperCollins Publishers; 2007.

CSA Survey Results Fall 2017			
Questions	**Answers**		
1	**State & County of residence**	New York	Saratoga	47
		New York	Albany	12
		New York	Rensselaer	5
		New York	Fulton	3
		New York	Montgomery	3
		New York	Schenectady	3
		New York	Warren	2
		New York	New York County	2
		New York	Livingston	2
		New York	Steuben	1
		New York	Not Disclosed	8
		Arizona	Maricopa	5
		Arizona	Phoenix	1
		Arizona	Not Disclosed	1
		California	Los Angeles	1
		Nevada	Clark	1
		Oklahoma	Tulsa	1
		Oregon	Multnomah	1
		Vermont	Addison	1
		Missouri	Howell	1
		Not Disclosed	Not Disclosed	1
	Total Participants			**102**

2	**Are you familiar or have you participated in a CSA?**	Yes - 36	No - 66	

2.2	**If YES to Q2, do you intend on participating again in the future?**	Yes - 20	No - 15	Not Disclosed 1

2.3	If No to Q2.2, What reasons would keep you from participating in the future? *Note - Some participants listed multiple reasons for not wanting to be part of a CSA.*	Cost	2
		Convenience	1
		Alternatives - have a garden, regularly shop at farmers market	7
		Time involved	1
		Too much food	6
		Unaware of local options	1
		Lack of variety/Lack of food choices	2
		Not Disclosed	4
3	What is most important to you when shopping? *Note - Some participants listed multiple food choice priorities.*	Price	39
		Organic	12
		Location of Origin	13
		Nutritional Value	36
		Preparations Requirements	10
		Appearance	13
4	How much do you spend on fruits & vegetables a week?	$0 - $10	12
		$11 - $20	22
		$21 - $30	18
		$31 - $40	6
		$41 - $50	21
		$51 - $60	4
		$61 - $70	1
		$70+	2
		Unknown	16

5	How much do you spend on groceries total per week?	This information was used to calculate the percentage of total budget devoted to fresh food.	0% - 10%	7
			11% - 20%	22
			21% - 30%	25
			31% - 40%	15
			41% - 50%	11
			51% - 60%	3
			61% - 70%	2
			81% - 90%	1
			Unknown	16
6	How many servings of "whole foods" do you eat a day? Description of what a "whole food" is was provided.		0 Servings	5
			1 Serving	11
			2 Servings	18
			3 Servings	21
			4 Servings	6
			5 Servings	11
			6 Servings	10
			7 Servings	3
			8 Servings	3
			9 Servings	1
			10 Servings	4
			11 Servings	1
			15+ Servings	5
			Not Disclosed	3
7	Do you think purchasing fresh healthy organic whole food is more expensive?		Yes	91
			No	11
8	Would you be surprised to find that there are many less expensive options for purchasing healthy organic whole foods than your average grocery store?		Yes	75
			No	26
			Not Disclosed	1

9	Do you think fresh whole foods are more nutritious and can help you avoid mental and physical sickness and disease? *Multiple choice.*	I think there is a direct link between food and mental/physical health, and I specifically choose fresh whole foods whenever possible.	61
		I think there is no connection between the food I eat and my mental and physical health.	6
		I think there is a connection between food and mental/physical health, but I don't base my food choices on that factor.	35
10	How much impact do you think you have on your local and regional economy with the choices you make when shopping for both food and non-food items? *Multiple choice.*	I do not think my choices have any effect on my local and regional economy.	4
		I know my choices have an effect on my local and regional economy, and I try to shop locally when it is convenient.	63
		I think my choices have a very small effect on my local and regional economy.	14
		I make every effort to shop locally and encourage others to do so, because I understand that consumerism drives the choices we are given as well as the health of my local and regional economy.	21
11	Do you enjoy cooking and tasting new foods, or simply eat the basics to survive?	Enjoy exploring new recipes and flavors	68
		Stick to the basics	34
12	With relation to question 11, does your level of cooking skills influence your choice in food?	Yes	62
		No	40

13	Does your household group make it a point to eat at least one meal together a week?	Yes	82
		No	13
		Single person household	7
14	Were you raised in a home where food was a significant part of your heritage/culture?	Yes	58
		No	44
15	Do you carry on those traditions or any new traditions in your own home today?	Yes	57
		No	45
16	Does anyone in your household have any special dietary preferences/concerns? (Select all that apply)	Vegetarian	12
		Vegan	2
		Gluten Free	11
		Allergies	25
		Religious/Cultural	8
		Diabetic/Reduced Sugar	7
		Paleo (High Protein/No sugar or carbs)	3
		No Pork	1
		Gout related	1
		Medical restrictions	1
		Reduced Salt	1
		Dairy intolerance	2
		None	54
17	Do you think the USA is progressing towards a "2 tier food system" where the cost fresh healthy whole food is becoming so expensive that only more affluent families can afford it, while those with lower incomes are limited to processed food choices that are less nutritious?	Yes	85
		No	17

18	Have you heard the term "Food Desert" meaning an area where fresh healthy whole food options are almost non-existent, regardless of cost?	Yes	53
		No	49

19	If you were told that the cost of food is inflated due to common practices like exporting a product such as potatoes to a foreign country, then importing potatoes from yet another country to stock grocery store shelves as a means to cycle funds to commercial agriculture companies and shipping companies, would you be surprised?	Yes	18
		No	84

20	How concerned are you about food security?	I am not concerned at all, there is plenty of food and little difference nutritionally between fresh whole foods and processed foods.	3
		I am concerned that changing weather patterns could affect the availability, both cost and quantity, of certain items (like a late frost killing off tender fruit trees buds).	33
		I think there is a difference nutritionally between fresh whole foods and processed foods, and I think some fresh whole food options are limited due to cost, but not quantity.	30
		I think that commercial agriculture's focus on high volume staple products and use of unsustainable farming methods could create a scenario where food security is an issue for everyone in the near future.	36

21	After taking this survey, are you interested in learning more about ways that you can save money, support local fresh food growers, improve your health, and influence your local and regional economy?	I am not interested, I think this survey was a waste of time.	1
		I am interested, some of these questions caused me to really think about the choices I make.	71
		I am interested, but I likely won't follow through if information is provided.	30